Using Google Earth in Libraries

PRACTICAL GUIDES FOR LIBRARIANS

About the Series

This innovative series written and edited for librarians by librarians provides authoritative, practical information and guidance on a wide spectrum of library processes and operations.

Books in the series are focused, describing practical and innovative solutions to a problem facing today's librarian and delivering step-by-step guidance for planning, creating, implementing, managing, and evaluating a wide range of services and programs.

The books are aimed at beginning and intermediate librarians needing basic instruction/guidance in a specific subject and at experienced librarians who need to gain knowledge in a new area or guidance in implementing a new program/service.

About the Series Editor

The **Practical Guides for Librarians** series was conceived by and is edited by M. Sandra Wood, MLS, MBA, AHIP, FMLA, Librarian Emerita, Penn State University Libraries.

M. Sandra Wood was a librarian at the George T. Harrell Library, The Milton S. Hershey Medical Center, College of Medicine, Pennsylvania State University, Hershey, PA, for over 35 years, specializing in reference, educational, and database services. Ms. Wood worked for several years as a Development Editor for Neal-Schuman Publishers.

Ms. Wood received a MLS from Indiana University and a MBA from the University of Maryland. She is a Fellow of the Medical Library Association and served as a member of MLA's Board of Directors from 1991 to 1995. Ms. Wood is founding and current editor of *Medical Reference Services Quarterly*, now in its 35th volume. She also was founding editor of the *Journal of Consumer Health on the Internet* and the *Journal of Electronic Resources in Medical Libraries* and served as editor/co-editor of both journals through 2011.

Titles in the Series

1. *How to Teach: A Practical Guide for Librarians* by Beverley E. Crane
2. *Implementing an Inclusive Staffing Model for Today's Reference Services* by Julia K. Nims, Paula Storm, and Robert Stevens
3. *Managing Digital Audiovisual Resources: A Practical Guide for Librarians* by Matthew C. Mariner
4. *Outsourcing Technology: A Practical Guide for Librarians* by Robin Hastings
5. *Making the Library Accessible for All: A Practical Guide for Librarians* by Jane Vincent
6. *Discovering and Using Historical Geographical Resources on the Web: A Practical Guide for Librarians* by Eva H. Dodsworth and L. W. Laliberté
7. *Digitization and Digital Archiving: A Practical Guide for Librarians* by Elizabeth R. Leggett
8. *Makerspaces: A Practical Guide for Librarians* by John J. Burke
9. *Implementing Web-Scale Discovery Services: A Practical Guide for Librarians* by JoLinda Thompson
10. *Using iPhones and iPads: A Practical Guide for Librarians* by Matthew Connolly and Tony Cosgrave
11. *Usability Testing: A Practical Guide for Librarians* by Rebecca Blakiston
12. *Mobile Devices: A Practical Guide for Librarians* by Ben Rawlins
13. *Going Beyond Loaning Books to Loaning Technologies: A Practical Guide for Librarians* by Janelle Sander, Lori S. Mestre, and Eric Kurt
14. *Children's Services Today: A Practical Guide for Librarians* by Jeanette Larson
15. *Genealogy: A Practical Guide for Librarians* by Katherine Pennavaria
16. *Collection Evaluation in Academic Libraries: A Practical Guide for Librarians* by Karen C. Kohn
17. *Creating Tutorials: A Practical Guide for Librarians* by Hannah Gascho Rempel and Maribeth Slebodnik
18. *Using Google Earth in Libraries: A Practical Guide for Librarians* by Eva H. Dodsworth and Andrew Nicholson

Using Google Earth in Libraries

A Practical Guide for Librarians

By Eva H. Dodsworth
and Andrew Nicholson

PRACTICAL GUIDES FOR LIBRARIANS, NO. 18

ROWMAN & LITTLEFIELD
Lanham • Boulder • New York • London

Published by Rowman & Littlefield
A wholly owned subsidiary of The Rowman & Littlefield Publishing Group, Inc.
4501 Forbes Boulevard, Suite 200, Lanham, Maryland 20706
www.rowman.com

Unit A, Whitacre Mews, 26-34 Stannary Street, London SE11 4AB

British Library Cataloguing in Publication Information Available

Library of Congress Cataloging-in-Publication Data
Dodsworth, Eva, 1976–
Using Google Earth in libraries : a practical guide for librarians / Eva Dodsworth, Andrew Nicholson.
pages cm. – (Practical guides for librarians ; no. 18)
Includes bibliographical references and index.
ISBN 978-1-4422-5503-6 (cloth : alk. paper) – ISBN 978-1-4422-5504-3 (pbk. : alk. paper) – ISBN 978-1-4422-5505-0 (ebook) 1. Information storage and retrieval systems–Geography. 2. Google Earth. 3. Geographic information systems. I. Nicholson, Andrew, 1970– II. Title.
Z699.5.G38D63 2015
910.285–dc23 2015016789

∞™ The paper used in this publication meets the minimum requirements of American National Standard for Information Sciences—Permanence of Paper for Printed Library Materials, ANSI/NISO Z39.48-1992.

Printed in the United States of America

Contents

Figures and Tables

Figures

Tables

Preface

June 2015 marked the tenth anniversary of Google Earth. Over the last decade since it was launched to the public as a free downloadable application, Google Earth has completely changed the way people use maps and geographical information. Today, Google Earth is so ubiquitous, it is difficult to imagine a time when there wasn't a mapping application to rely on for reference and research.

Prior to the 2005 launch of Google Earth, the world of online mapping was still very new terrain. Paper maps for tourism and recreation purposes were still the dominant format. Talk of electronic maps focused on those available on CD-ROM which were clunky to use, expensive, and not practical for the driver or hiker needing the information at their fingertips. On the Internet, MapQuest had been around for nine years and was seen by many as the "go to" source for online mapping, but had limited capabilities beyond providing routes, hotels, and restaurant information. It also didn't have satellite imagery and offered limited coverage outside of North America.

The launch of Google Earth (and its sister app Google Maps) would change all this by providing maps and satellite views of every part of the globe. For the first time, people could see in an easy-to-use, digital way, just how geographically connected everyone is. For example, Google Earth would provide the world with a virtual globe complete with clear satellite imagery and a host of embedded datasets containing geographical information from several categories including current issues, scientific research, historical maps, tourism and traffic data, and so much more. Google Earth will build on this success with the addition of its "Street View" imagery, allowing people to see for themselves complete streetscapes of their own cities, but also strange and exotic places that they had only imagined visiting before. Further enhancing the user experience of Google Earth is its interactive functionality, including its map making and mashup capabilities that would set Google Earth apart from other online mapping apps both before its launch and even since then. Arguably, it can be said that Google Earth has been one of the defining web tools of the past ten years for its ability to visualize user-created maps with real world data and then share it easily with whomever was interested.

This book is a guide to the Google Earth application, but with an extra emphasis given to its role in education and research. Not only can librarians and educators make their own maps with Google Earth, but users can also utilize the application to promote their own existing map and data collections; foster a culture of data sharing in their workplace; develop spatial literacy skills amongst their students; and empower learners of all ages to think about the world from different perspectives. Perhaps most importantly, *Using Google Earth in Libraries: A Practical Guide for Librarians* is not written just for geography librarians or earth science teachers, but for educators in a wide variety of disciplines and subjects. As readers will discover from exploring these pages, Google Earth has widespread applicability across subject disciplines and interests; and can be adopted by learners of all ages and backgrounds regardless of their experiences with maps, mapping technology, or computers in general.

Organization

Using Google Earth in Libraries: A Practical Guide for Librarians contains five chapters outlining how to make the most out of the Google Earth mapping application in the classroom. Along with providing step-by-step instructions for the most basic to the most advanced functions of the program, this guide also provides information about where to find Google Earth data (KML files) available online.

Chapter 1 serves up a general orientation and overview to the Google Earth application, its mapmaking tools, embedded layers, historical imagery, and some of the additional features found in the more advanced Google Earth Pro application. This chapter will help the first time and novice user wade into the Google Earth application, leading them to discover features and functions of the program that will both surprise and delight.

An overview of how others are using Google Earth is the focus of chapter 2 including some selected pathfinders to where a user can find Google Earth data (KML files) created and offered up by external sources, including academic libraries and government agencies. Many libraries have used Google Earth as a finding aid for their map and air photo collections, while others have gone even further in providing the images and data themselves in a "born digital" format for instant access in Google Earth. Many government agencies in the United States and Canada have also embraced the call for "open data," releasing publicly collected data in KML format for sharing in Google Earth.

Adopting Google Earth as a learning tool in a classroom is discussed in chapter 3. Lesson plans are reviewed which utilize the information presented in Google Earth but also the interactive map making functions which make Google Earth such a valuable technology in a learning plan. Examples from both K-12 and higher education are outlined as are case studies from the Sciences, Social Sciences, and the Humanities. Learning concepts such as "Bloom's Taxonomy," "Blended Learning," and "Place Based Learning" are also touched on in the context of bringing Google Earth into the classroom.

Chapter 4 will take the reader "under the hood" of Google Earth and give greater detail in how to take full advantage of Google Earth in customizing features and scripting KML files. This includes creating animated Google Earth data, producing professional placemark balloons, and utilizing file conversion tools that will allow for the importation of other geospatial datasets into Google Earth. Advanced functions such as georeferencing of external maps and imagery and geocoding of address data are also described, while providing recommendations on the best ways to share Google Earth data.

Getting "Hands on" with Google Earth is the focus of chapter 5, as readers are provided with a selection of tutorials designed to train users in how to perform specific tasks such as creating placemarks, georeferencing, creating paths and tours, and working with Google Fusion tables.

Lastly, Google Earth is a fun and engaging tool to spend time with. Whether for serious learning, or chilling out by planning a dream vacation, Google Earth has something to offer to virtually anyone who has the time and a computer with an Internet connection. Users easily lose themselves in the imagery and maps, but at the same time feel empowered by making their own maps and visualizations to share with others.

CHAPTER 1

Google Earth: A World Resource for Research

IN THIS CHAPTER

- ▷ Google Earth: Orientation to the Interface
- ▷ General Map Making Tools
- ▷ Layers
- ▷ Historical Imagery
- ▷ Google Street View
- ▷ Google Earth Pro: Additional Features

GOOGLE EARTH IS a free online three-dimensional virtual globe that has been growing yearly in popularity as an exploration, visualization, and research tool. With its interactive display, freely available images and 3D views, built-in sources of geographic information and mapping capabilities, educators and researchers are using Google Earth to explore relationships among datasets, including the ability to project, model, and present research data in new ways. Users can view millions of stitched images of the globe overlaid over landscapes representing hills, valleys, and water. Google Earth appeals to users in many areas of research, and with varying expertise levels. With its reliance on Keyhole Markup Language (KML) files which are easy to create and modify, Google Earth allows users to embrace a Geographic Information System (GIS) technology without the need for specialized knowledge in GIS, cartography, or programming.

This chapter describes the essential elements and features of the Google Earth Mapping application that make it a valuable resource for libraries and educators. Since its release to the public in 2005, some writers have labeled Google Earth as a "liberation technology" (Hong, 2013). Perhaps this is an overstatement, but there is no doubting the power and utility of Google Earth, especially in that it is a rich resource of geographic and cultural information.

Background

With Google Earth, the whole world is at the user's fingertips, from place finders, to route creations, to virtual visits from coast to coast, users can explore, visualize, and learn about areas they have never been to. Michael Jones, the chief technologist for Google Earth, stated, "We thought it would be great if everyday people could use this spatial knowledge to make better decisions. … So we set out to build an Earth browser to help people better understand their planet" (Wagner, 2006).

For the novice user, Google Earth may first appear to be an air photo and satellite image portal; however, the partnerships between Google Earth and various organizations have built this product up to be much more than that. There are various built-in features within the program that shed light on community growth and environmental changes. With historical imagery available, users can track geographical changes over time and study factors like population growth, urban re-developments, and environmental changes. With supported importation tools, Google Earth becomes transformed into a visualization and mapping program, enabling users to add personal and customized layers to build, study, and display the information they need.

Google Earth was originally named EarthViewer 3D, created by Keyhole, Inc., a company acquired by Google in 2004. Keyhole, Inc., first developed their virtual globe in 2001, a project partly funded by the Central Intelligence Agency (CIA); however, it wasn't until Google acquired the company and made the virtual globe technology publically available by download at no cost that it became an international sensation (Tuttle, Anderson, and Huff, 2008).

Google Earth started off as an application that modelled the Earth by simply using current satellite imagery and aerial photography. Many areas of the world were captured with images using satellites with a resolution of about 15 m per pixel, allowing users to define buildings and terrain features. Users were seeing parts of the world they had never seen before. Researchers likewise were exploring virtually, examining the world like a newborn for the first time. Upon Google Earth's release, new scientific discoveries emerged. Forests never known about were spotted, and endangered species never recorded before were discovered. One such example is the discovery of Mozambique's Mount Mabu, a mountain that was unmapped, unexplored, and unknown to anybody outside of the locals inhabiting it. The scientist, Dr. Bayliss, who discovered it virtually, paid many visits to this mountain and with his team discovered rare birds, a new pygmy chameleon, a previously unknown bush viper and two new species of snakes, four new butterfly species, a crab, and five plants (GrrrlScientist, 2013).

In 2006, 37 percent of the world's population was covered by high-resolution imagery. By 2012, coverage increased to 75 percent of the global population (King, 2012). Today, Google continues to update a lot of the coverage with 2.5 m SpotImage satellite imagery (Wikipedia, 2014). Many metropolitan areas are also covered by aerial photography with resolution as high as a few cm per pixel, allowing users to make out vehicles and even individuals. These images are acquired from a variety of sources, but mainly from city or state/provincial governments.

Some users may be under the false assumption that the imagery seen in Google Earth is live. The images are taken at different times, during different cloud covers and are generally updated every three years. After a few years of updating images, Google Earth started developing an archive of images and released them to the public in 2009 as part of their new feature, the time slider. With the time slider, users can slide or toggle between two or more years to compare the changes that the passing of time may have made. Some areas will have more imagery updates than others, especially in spots that have had drastic changes occur to the landscape, due to natural disasters such as tsunamis, floods, hurricanes, and fires.

Google Earth was originally available with three different licenses, Google Earth (a free version), Google Earth Pro ($400 per year), and Google Earth Plus which has been since discontinued. In early 2015, Google Earth Pro became freely available to the public. As the downloads of the regular Google Earth application far outweigh those of Google Earth Pro, this book focuses on the regular application, although Google Earth Pro will also be briefly discussed later in this chapter. Google Earth can be used on any Windows, Macintosh, Linux kernel, and FreeBSD operating systems. There is also a mobile version of Google Earth available for Android OS and iOS (iPad/iPhone). You can download the most recent version of Google Earth from <http://www.google.com/earth/> (As of May 2014, Version 7.1.2 of Google Earth is available).

Google Earth: Orientation for First-Time Users

Along with its free availability, the most appealing aspect of Google Earth for research and mapping is its ease of use and low learning curve. One of the inspirations to write this book has come from watching first-time users and geography neophytes embrace Google Earth as a tool for their own learning and research. These users have ranged from first-year university students to middle-aged library staff, each of whom discovered something in Google Earth that they could take away.

Upon installing and opening Google Earth, you are presented immediately with a Start-Up Tip window with the Earth as a dazzling blue and green globe floating in space behind it (figure 1.1). As with many software products, particularly from Windows, the

Figure 1.1. Google Earth as it appears when first opened.
Source: Google Earth 7.1: <http://earth.google.com/>. Map Data: US Dept of State Geographer, 2014 Google, 2009 GeoBasis – DE/BKG.

Start-Up Tips are set as a default option to help first-time users get orientated to the program. On the left of the Start Tip window is a short menu of links to tip information on performing particular tasks from "Navigating in Google Earth" and "Searching for places" to "Using tours" and "Viewing the past." Each of these links takes the user to Google Earth Help webpages, and short tutorial videos. Once you are comfortable using Google Earth, you can switch off the Tips window, by unchecking the "Show tips at start-up" box. Once you are finished going through the Tips, or you have turned them off as a default, you will then be presented with the Google Earth Interface screen. Depending on where you are located at that moment, Google Earth will automatically turn its digital globe to your home country.

The Google Earth interface is made up of the main Google Earth viewing window and the sidebar menu on the left side of the screen, which contains smaller windows for users to access "Search," "Places," and "Layers." There are also navigation tools and a top menu bar providing access to a variety of tools.

Search

The Search box, located on the top left of the Google Earth screen allows users to search for any place located in the Google Earth database. This includes not just geographical place names, but also street addresses, geographic coordinates, businesses, parks, and many other places of interest. Included is the "Get Directions" search functionality, popular in Google Maps, as well as a "History" option allowing users to look back and reference previous search histories.

Places

The Places window contains two distinct sections which are key to understanding how to organize your files and successfully manage map projects in Google Earth. The top section is called "My Places" and this is where you will keep your KML and KMZ files that you will either create or wish to keep whenever you reopen Google Earth on that computer. The second section is called "Temporary Places" which is where KML files from other sources are first placed when they are opened from an external website or from your own computer. Anything that is in the Temporary Places section when Google Earth is closed is removed from the application, while KML files that are placed under "My Places" are retained inside the Google Earth interface and can be easily returned to when Google Earth is opened again. To move KML data from Temporary Places to My Places, you can always "drag and drop" the files or use the "Save to My Places" option found under the File menu.

Layers

The Layers window on the bottom left of the screen contains the embedded layers that are part of the Google Earth database and can be switched on or off depending on what you want to see. It is in the Layers box, for example, that you can control the visibility of national and municipal boundaries (Borders and Labels category). The Roads layer is also available here. As with each of these windows, you can use the gray triangle to open up the options under each layer category.

The Layers window also contains many specialized KML collections created by outside groups, however they have now been made available for Google Earth users. Exploring the "Gallery" and the "Global Awareness" sections will highlight many of these interesting KML collections, some of which may be of particular interest to researchers. These are discussed in more detail later in the chapter.

Navigation

The main navigation tools used in Google Earth are located in the top right window (figure 1.2). The top dial (with the eye) allows you to alter the tilt and aspect of your view, which is especially useful for doing fly-thrus and looking at 3D models. The dial below

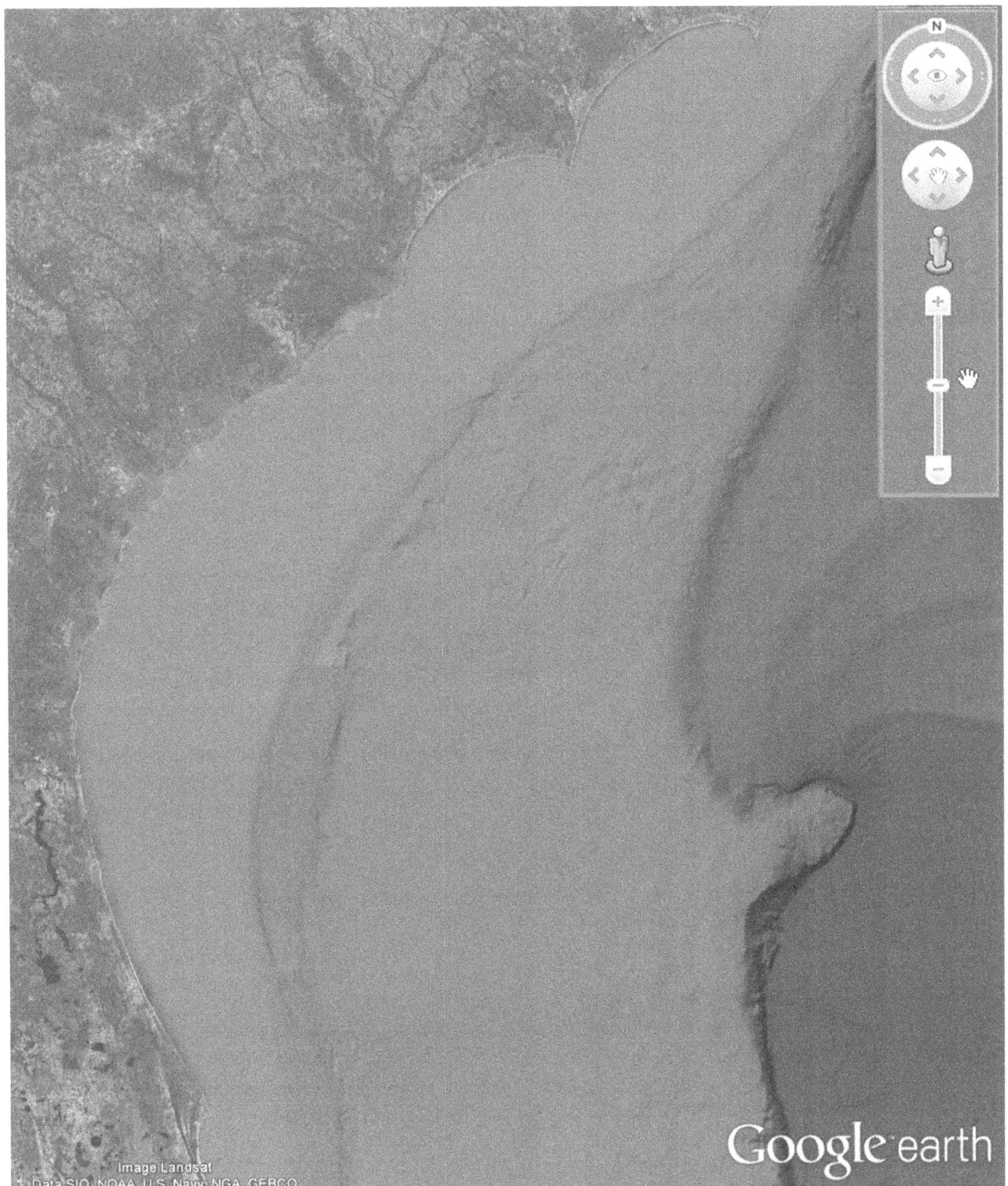

Figure 1.2. Google Earth Navigation Tools.
Source: Google Earth 7.1: <http://earth.google.com/>, Source: Landsat

is your direction control and underneath is your zoom slider allowing you to zoom in or out of your Google Earth view.

Zooming in closer to the ground will bring up the Street View character icon known as the "Pegman" between the dials and the slider. When clicking on the "Pegman" and dragging him over your Google View perspective, streets with available Street View imagery will become highlighted, and you can then drop the "Pegman" onto a street to bring up the Street View imagery.

Google Earth Top or App Menu

As with many Windows or Mac Applications, the Top or App menu appears across the top of the screen with familiar headings such as File, Edit, and others (table 1.1). In Google Earth, those headings are also available as drop down menus (figure 1.3).

Located at the top of the Google Earth View Screen, the "Action Menu" bar provides options for marking up the Google Earth imagery, sharing and printing the images, as well as measuring distances and adjusting for different perspectives from Sun placement to historical image overlays (figure 1.4).

General Map-Making Tools

Google Earth's straightforward user interface makes it an easy choice for users wishing to explore and discover their surroundings using satellite imagery, aerial photography,

Table 1.1. Google Earth Top Menu Options

APP MENU HEADING	SUB MENU -KEY ITEMS	DESCRIPTION
Google Earth	Preferences	Provides many options for customizing your Google Earth performance, such as 3D views, cache settings, Tour settings, and navigation.
File	Open, Save	Open or save your KMZ files; share them via email or Google Maps
Edit	Copy; Paste; Delete; Show Elevation Profile	Along with standard features, you look at a Elevation Profile of a Path (see page 10 for more on Elevation Profiles)
View	Toolbars, Sidebar, many other options	Many options available for altering your view; some are also available on the Action Toolbar
Tools	Ruler, GPS, Flight Simulator	Measure distances; interface and obtain points from a GPS unit
Add	Folder, Placemark, Polygon, and more	Add new folders and mark up your Google Earth project with your own layers; also available in the Action Menu Toolbar (see below)
Window	Minimize, Zoom	Adjusting the Google Earth view
Help	Help resources; Keyboard Shortcuts; Start-up Tips; and more	Learn Google Earth inside and out exploring the excellent Help menus

Figure 1.3. Google Earth "Action Menu" Bar.
Source: Google Earth 7.1: <http://earth.google.com/>, Source: 2015 Google, Landsat

Street View imagery, and the thousands of resources available through the Layers feature. Additionally, Google Earth allows the users to "mark up" the imagery, adding their own Google Earth layers to illustrate their research.

Before highlighting the Google Earth Map Making tools, it should be stressed that the key to successfully using Google Earth, particularly if you are creating your own maps, is to clearly organize your KML files and save them often. The best way to do this is to organize your data into a project *Folder*. When beginning a new Google Earth project, it is highly recommended that you "Add a Folder" to the "Places" window, especially in the "My Places" section as your first step. Into this folder will go your *Placemarks*, *Polygons*, *Paths*, *Image Overlays*, and *Photos* that you will be adding to Google Earth. When your project is complete or you wish to share the files or move to another computer with Google Earth installed, you just need to use the "Save Place As" option found under the File Menu.

Basic Menu Items

Google Earth comes equipped with several built-in map-making tools. Perhaps the most popular and identifiable as a "Google" product is the *Placemark*, used to denote points on the map. Placemarks are also often used to arrange events or tell a story in chronological order, using numbers as the symbol of choice. Placemarks can be added without knowing actual addresses of points of interest. This greatly expands the ability to map locations as users often search the database not by address but by subject name, such as hotel or coffee shop. Be it the location of your house, your favorite travel destinations, or a display

MENU TOOLS	FUNCTION
	Show/Hide Sidebar
	Add Placemark
	Add Polygon
	Add Path
	Add Image Overlay
	Record a Tour
	Show Historical Imagery
	Show Sunlight across the Landscape
	Switch between Earth, Sky, and other planets
	Show Ruler
	Email
	Print
	Save Image
	View in Google Maps

Figure 1.4. Google Earth "Action Menu" Options
Source: Google Earth 7.1,<http://earth.google.com/>.

of geocoded addresses, you can apply Placemarks using the default "yellow pushpin," or choose from a selection of Placemark symbols found in Google Earth. If you want to be truly creative, you can find a photo thumbnail or make your own symbol as a Placemark. For example, a researcher mapping locations of artifacts found in Ancient Pompeiian households may wish to use thumbnail pictures of each of the artifacts in their placemarks to better illustrate where in the Pompeiian homes the artifacts were found and what they tell us about the class and lifestyle of the people who lived in the homes.

Once a Placemark is added, users can label the Placemark and add information to the "balloon" or window that pops up when a user clicks on it. HTML code can be used to

WHAT ARE KML AND KMZ FILES?

KML (Keyhole Markup Language) file is the native Google Earth file format used for modeling and storing geographic features. All of the data embedded in Google Earth, and those that are created by users can be saved as KML files. A KMZ (Keyhole Markup Zip) file is a holder or basket of KML files, which you can save as a Google Earth project.

customize and organize the text, such as including tables, bullets, colored text with variable fonts and font sizes. URLs and images are often included in these pop-ups as well to provide users with as much information about each individual Placemark as possible. For example, if somebody was interested in creating a map of dog park locations within a state or province, each dog park location would be denoted by a Placemark, labeled appropriately with the dog park name. The Placemark description could include hours of operation, services such as water fountain or plastic bags, rules and regulations, an image of the park, and a link to the city website that maintains the park.

Placemarks work really well when trying to tag a specific location. Users however may be interested in highlighting and describing features, such as a building or a forest. Instead of a point, users can draw a polygon to highlight an area in the imagery. Google Earth users can add a *Polygon* layer which can be drawn on the Google Earth imagery and can be of different colors or even made transparent in their center or fill, with only their outline showing. For example, a CSI Forensics Investigator may wish to lay out the perimeter of a crime scene. The investigator can add (or more accurately draw) a Polygon around where evidence is located and where the victim(s) were found. The Polygon can cover buildings, cars, fields, and whole properties and then be highlighted with colors (or made transparent to make the inside visible and outline clearly distinguishable). A description of what can be found in the crime scene can be added to the Polygon, and the layer can then be shared with fellow investigators, police, and the media.

Paths are ideal for Google Earth users looking to map out a route or a tour through a neighborhood, a city, or across a country. The Paths will appear as a line in Google Earth in which you can change the color and thickness. You can then save your Path as a *Tour* even adding audio and photos and converting to a video file for playback. For example, Paths and Tours can be part of a presentation project in which history students can present their research findings on a "new world" explorer and can map out their route across the oceans and make Paths of discovery through the uncharted mass of the new world. Students can present this video tour with Google Earth, including photos and links to research related to the explorer.

One of the most useful elements of Google Earth is the ability to measure distances from one place to another, from a straight line to around several corners. The *Ruler* button is located in the Google Action Menu Toolbar as well as under the "Tools" Menu. In Google Earth you only have the ability to measure Lines and Paths and then save them to your project. In Google Earth Pro, you also have the ability to measure Polygons, Circles, as well as 3D paths and 3D polygons.

Importing External Files

The flexibility and customization available within Google Earth allows users to visualize and manipulate not only data available within the program, but their own personal data by importing images, photos, and KML files. Image Overlays (or "map overlays" as they are sometimes referred) allow the user to add and overlay another map onto the Google Earth imagery and "georeference" it into place. The process is remarkably easy as long as you can see familiar geographic features between the Google Earth imagery and your map overlay. You can move, rotate, or adjust your map overlay image to make it more opaque for a more accurate georeference. Image Overlays are an ideal tool and can make historical research especially interesting, comparing landscape or urban change over time. Overlays have proven to be critical in research when studying circumstances surrounding historical events like battlefields or natural changes. For example comparing the changes to the Toronto waterfront over hundreds of years can be done by taking scans of maps and overlaying them on top of each other in Google Earth. With each new Image Overlay, now a Google Earth layer, you can switch them on and off and compare each Overlay with current Google Earth imagery. You can also draw Polygons or add Placemarks to note changes in shoreline or features that existed at one time, but have changed or disappeared.

If users wish to add photographs into Google Earth, but do not wish to overlay them on top of the Earth's surface, they can use the *Add Photo* feature, allowing them to select a photo from their local computer drive. Users can then place one or multiple photos anywhere they wish to in Google Earth. Alternatively, photos can also be hosted on an external website which you can link and embed directly into a Placemark. When users click on the Placemark, the photo will then appear in the pop-up window.

Another option for including Photos in Google Earth is to utilize Panoramio, a Google owned geolocation-oriented photo-sharing website. Panoramio is available from within the *Photos* Layers list, but users can upload their images directly to Panoramio's website, <http://www.panoramio.com/>, and while there, geotag the images to the approximate location the photo was taken. Once geotagged, when the photos are added into Google Earth, the image will automatically be placed in the correct geographic location.

Outside of photo and image importation, many users are able to use Google Earth as a mapping program by using the KML importation functionality. Files that are saved with the extension .kml or .kmz are able to be imported into the program. These files can be Placemarks, Paths, or Polygons and can be as simple as having a label denoting what it is, or as complex as having built-in dynamic pop-up windows, screen overlays, animation, video, sound, and much more. Chapter 4 will discuss KML files in much more detail. With Google Earth becoming more popular as a means of doing research, increasing numbers of people and groups are creating their own KML files and making them available online to share with others. Google has recognized this trend of KML creations and customizations and included a *Network Link* function to allow users to add these online hosted KML files. Rather than downloading the KML file onto your computer or USB key, Google Earth can pull the KML files directly from the host website right into your Google Earth window. By placing the link to the KML files into the Network Link window, a new layer of KML will be added to your existing set of KML layers.

KML files can be created by writing the code in a text program, by copying the code using an online tool, or by converting them from other file formats such as Esri's shapefile. Conversions, translations, and KML creations will also be discussed in detail in chapter 4.

Elevation Profile

Along with aerial and satellite imagery available, many researchers have also found the Elevation Profile tool to be especially useful. The elevation profile can be activated by adding a Path, and then selecting "Show Elevation Profile" from the Path Menu options. The Elevation Profile cross-section screen will then appear at the bottom of the Google Earth window. Moving the cursor along the profile will also indicate where on the path a particular elevation value is located.

As expected, geographers have found the Elevation Profile feature in Google Earth especially useful. Researchers can now visualize a cross-section of the Earth's terrain in minutes by drawing a path across a landscape. Google Earth and the Elevation Profile were in fact a large part of a study looking at drainage in the Susquehanna River Basin in Pennsylvania (figure 1.5). So much so, the researcher even included her KML files with her research paper for public download and further study (Lee, 2013).

The use of the Elevation Profile is also discussed by several authors at length in the edited volume, *Google Earth and Virtual Visualizations in Geoscience and Research*. Perhaps the most interesting and surprising use of the Elevation Profile in the book is the chapter on Benedict Arnold's march to Quebec City in 1775, in which the authors retrace the steps of the American revolutionary general and his failed expedition to capture the British colony of Quebec. The expedition was a disaster in large part owing to the difficult geography of northern Maine and inaccurate maps of the area. Only half the number of men who began the expedition made it to the edge of Quebec City and even those who did make it, were too exhausted and starving to mount a serious challenge to British forces. The difficulties of navigating across two steep river basins and a tangle of rocky streams and rivers is highlighted by visualizing the route in Google Earth and its Elevation Profile feature (Ruegar and Beck, 2012).

Google Earth Projections and Grids

A frequent question users ask of Google Earth, especially from cartographers, is what projection does Google Earth use? By projection, they are asking how the spherical nature of the Earth can be represented on a flat screen with minimal distortion in area, size, and distances. Every map and digital map of the Earth's surface ever created has struggled with this problem, and geographers get by it by picking a particular projection method to stretch the Earth and map it based on their specific needs.

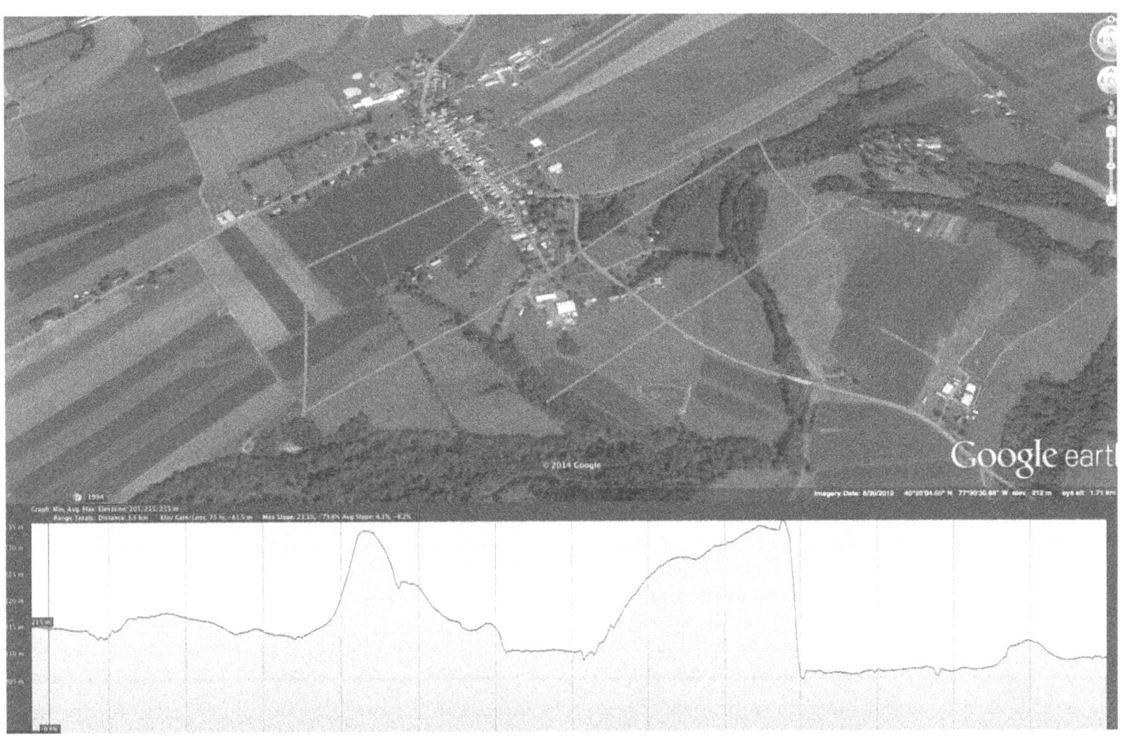

Figure 1.5. A Susquehanna profile line displayed in Google Earth.
Source: Google Earth 7.1, <http://earth.google.com/>. Map Data: 2014 Google

Google Earth uses a "Simple Cylindrical" projection similar to orthographic projection, and assumes a position that we are looking at the Earth from an orbiting satellite or platform high above the surface. Even though the projection is set in Google Earth and cannot be changed, you do have the ability to change the "Lat/Log" coordinate system being presented on the screen. The default is "Degrees, Minutes, Seconds," but options also include Universal Transverse Mercator (UTM), Decimal Degrees, and Degrees, Decimal Minutes. You can change these in the Google Earth Preferences/Options Window. You can also visualize the coordinates as a Grid on Google Earth, through the View . . . Grid menu option.

GPS Devices and Google Earth

GPS (Global Positioning Systems) devices are now in common use by almost everyone whether they are conducting field research or are looking for their favorite coffee shop in a strange city. Almost every mobile device has some kind of GPS app or capability of location tracking; however, their accuracy and functionality have limitations depending on the device. For people who need to rely on GPS coordinates for safe hiking or research data collection in the field, they will typically obtain a dedicated GPS device (such as those manufactured by Magellan or Garmin), that allows not just for navigation but also for accurate data and geographic coordinate collection allowing the user to record research findings outside with matching geographic and elevation readings. These findings can be recorded as "waypoints" which will be a specific location, or as "tracks" which might be a route or trail undertaken as part of the research. Once you have collected your research data with such a GPS device, you can connect your device to your computer and import your GPS data directly onto Google Earth. "Waypoint" data will typically appear as Placemarks in Google Earth while "Tracks" will appear as Paths. With your GPS synced with Google Earth, you can then assess the accuracy of your waypoint locations and save them as KML files.

Researchers involved in field studies using GPS devices have written extensively about how Google Earth has helped extend their research into new areas. For example archaeologists working in the United States and Cyprus used Google Earth to first explore sensitive areas that may be harmed by a full-on archaeological excavation. They used GPS devices to record reference points for a geophysical survey that they could then use to determine if an area was suitable for archeological study (Williams, Baker, and Ault, 2012).

Layers

Google Earth Outreach enables any organization to quickly and easily get the resources it needs to create compelling stories through Google Earth layers. The program includes comprehensive online guides, video tutorials, and case studies about using Google Earth specifically targeted to the needs of non-profit organizations. In addition, there are online forums connecting new participants to Global Awareness partners and experienced programmers who can assist in developing Keyhole Markup Language (KML) layers for Google Earth. These forums, actively moderated by Google Earth Outreach staff, serve to foster discussion and cooperation among organizations and the broader Google Earth community.

Google Earth enables individuals and organizations to easily retrieve the resources they need through guides, video stories, photographs, and case studies. The Layers feature is a dynamic and vital application of Google Earth that provides a wide variety of themes that can be displayed over the map. Several third-party non-profit groups and organizations are included in the Google Earth database, typically environmental or humanitarian in nature, providing related information to raise awareness about people, animals, nature, and the Earth. These user-friendly layers consist of points, lines, and polygons covering various geographical areas, and quite often provide links to more data, information, and resources. Some popular layers include Weather, 3D Buildings, Terrain, Rumsey Historical Maps, Panoramio images, 360 Cities, and the National Geographic Magazine. What is intriguing about the content of the layers is that many of them are actually updated periodically. Because Google Earth requires an Internet connection, users are always connecting to the latest layers. In fact, some layers like Webcam, Traffic, and Weather provide live data stream into Google Earth. The Weather layer for example offers current cloud display, radar, and conditions and forecasts. So with a click of the button, users can monitor the current temperature outside while map-making inside.

The layers are accessible from the lower left-hand side of Google Earth. Several are organized into folders and can be expanded by clicking on the triangle next to the folder. Once a layer is activated, it is immediately seen on the map. Turning on all of the layers at the same time will cover the map view, so it is best to view a few layers at a time.

Base Layers

The themes available in the list of layers vary quite a bit from one another. Some are very general, like Terrain, Roads, Borders, and Labels, but they are very important nonetheless. Having roads, for example, overlaid on top of aerial photography dramatically enhances the search and discovery experience. Since people are more accustomed to looking at a road map than aerial photograph, roads with names labeled assists with orientation and identification. This is particularly useful when using the time slider feature to examine older aerial photos. The roads include freeways, toll ways, major and minor roads, and even some pedestrian walkways and private roads.

Borders and Labels assist with navigational awareness as the political boundaries help make sense of the seamless imagery. Labels are particularly useful as they include large cities, small towns, and everything in between.

Perhaps one of the most well knows layers is "Places," which includes a variety of places of interests like banks, daycares, restaurants, bars, stores, lodging, entertainment, shopping, recreation, and so forth. Users can zoom into their area of interest, activate the "Places" layer, and see all the establishments as individual symbolized Placemarks.

3D Buildings

On the opposite end of the spectrum, there are layers that do not represent the world as a collection of points, but rather as three-dimensional realistic models. In the almost ten years since its introduction, new 3D terrain landscapes and Model buildings have been added to the application. Inside the Layers window, users will find the "3D Buildings" category, including Models made from Photorealistic data and a separate 3D Trees layer. This layer includes numerous famous sites such as the Empire State Building, the Eiffel Tower, the Taj Mahal, to name a few, but in the most recent years entire cities have been built in 3D. Universities, colleges, high schools, high-rises, theatres, businesses, and others have been built by the users themselves. Users can create their own 3D models using modeling applications such as SketchUp and Google Building Maker or they can find ready-built models on websites such as the Sketchup Warehouse, which can be found at <https://3dwarehouse.sketchup.com>.

More recently, however, Google has been improving the resolution of the 3D terrain around the world. Places like Mount St. Helens, Stonehenge, Big Ben, and Disney World are available for viewing in high resolution.

Gallery

All of the layers mentioned above are great tools for becoming familiar with a geographical area. Users typically use the general layers for planning visits or trips or to learn about or discover an area without actually visiting it. There are layers, however, that are more theme-oriented and provide a deeper understanding of the world. Many of these are within the Gallery folder, a comprehensive compilation of photo-related layers available from a number of organizations.

Rumsey Historical Maps Layer

One of the most popular layers within the Gallery folder is well known among cartographers, the Rumsey Historical Maps layer. This layer consists of over 150 historical maps, which is just a small sampling of David Rumsey's personal collection of 150,000 maps dating from 1680 to 1930. David Rumsey is a map collector based in California and has digitized and shared 50,000 maps on his website <http://davidrumsey.com>. His map collection focuses on rare eighteenth and nineteenth-century North American and South American maps (Cartography Associates, 2014). Each of the maps in the Rumsey Historical Maps Layer have been georeferenced and drape the earth's surface when activated. Historical maps tell a story on their own, but when compared to the modern landscape, a lot can be said about the passage of time and its effects on natural features.

Wikiloc

The Wikiloc layer is a composition of millions of trails that users are continuously adding to and sharing with the world. Users can add their favorite GPS routes, including the elevation via the Wikiloc website, <wikiloc.com>. The trails that are available from this website can be discovered and mapped through Google Earth's Wikiloc layer. The placemark icons indicate the type of trail, such as hiking, biking, skiing, as well as the trail route, length, elevation, and whether the trip is one-way or round trip. The trail difficulty is also described by the contributor. There is something for everyone as the trails vary in length and difficulty. Some are leisurely walks around town and others are pathways up into the mountains.

Wikiloc was created in 2006 as a hobby by Catalonian Information Engineer and mountain climber Jordi Ramot. He would draw his routes on Google Maps and realized that he could be sharing them with others. In 2008 Google Earth embedded his Wikiloc project into the Gallery layer (Kiremex, 2013).

360Cities

360Cities is another Gallery layer item that offers a sampling of the web's largest collection of georeferenced, interactive panoramic photos. 360Cities <http://www.360cities

.net/>, established in 2007 has a vast database of imagery thanks to a growing community of contributors. These include panoramic interior and exterior shots that have been added to the 360Cities website or a related Google Earth layer. Images have been taken all over the world, from historical cities like Pompeii and Rome, to popular landmarks like the Eiffel Tower to remote areas in caves and on beaches on islands like Porto Santo, totaling approximately 10,000 individual photographs. 360Cities allows users to explore areas of the world that they have never been to, inviting them inside buildings and natural enclosures. Used as a cultural exploration tool, many users can virtually fly into towns they have never heard of and explore on their own.

Global Awareness

Global Awareness is a compilation of several layers that relate to concerns around the environment, humanitarianism, cultural and educational issues, as well as disaster relief/response. Many non-profit groups, government agencies, scientists, and citizens have contributed information to develop and share this collection of Google Earth layers. From fair trade coffee production sites, to locations of Earthwatch expeditions, to locations of endangered species, some layers are more in depth than others, such as the United States Holocaust Memorial Museum, the Crisis in Darfur, the Atlas of Our Changing Environment from the United Nations, and finally the Jane Goodall Institute: Gombe Chimpanzee Blog.

United States Holocaust Memorial Museum: Crisis in Darfur

In 2007, the United States Holocaust Memorial Museum in Washington, D.C. partnered with Google Earth to create a layer entitled "USHMM: Crisis in Darfur," a resource that informs the public about the damage and level of destruction done to villages, refugee camps, and communities in Darfur. In 2003, Sudanese militants and soldiers attacked the people of Sudan, driving over 2.5 million people away from their homes. In 2004, a genocide emergency was declared for Darfur, Sudan (United States Holocaust Memorial Museum, 2009). This layer raises awareness about genocide and encourages and inspires users to help prevent a similar situation from occurring (Moore, 2007).

The layer consists of photographs, data, and eyewitness testimony, and the imagery available for this region, specifically for Sudan and Chad, are high resolution allowing users to view the more than 1,600 damaged homes and destroyed villages. The remnants of larger communities, including schools and mosques are also visible (Moore, 2007).

United Nations Environment Programme: Atlas of Our Changing Environment

A partnership between the United Nations Environment Programme (UNEP) and Google Earth brings users the "UNEP: Atlas of our Changing Environment," which includes image overlays of environmental hotspots from around the world. Through the Google Earth imagery, users can see environmental changes on the planet including retreating glaciers, melting snow from mountains, deforestation, and forest fires; by comparing photos taken up to thirty-five years ago with more recent photos, users may be inspired to help promote not only awareness, but change as well.

UNEP's partnership with Google was built upon the success of the UNEP publication *One Planet, Many People: Atlas of Our Changing Environment*, a collaboration with

UNEP, NASA, USGS, and the University of Maryland. One hundred locations depicted in the printed version were made available digitally in Google Earth in 2006. After several years, this layer is still active, is regularly updated with more imagery, and continues to be used by many to study and research the effects of man on the environment.

Jane Goodall Institute: Gombe Chimpanzee Blog

The goal of the Jane Goodall Institute is to preserve African great apes and their habitats by conducting ongoing field research and conservation projects. Jane Goodall and her fellow researchers have been studying and observing chimpanzees in Tanzania for the past fifty years. However they were not only observing chimpanzee behavior but also their habitats and the occurring deforestation. According to Lilian Pintea, vice president of conservation science for the Jane Goodall Institute, 80 percent of the forests around the park being studied were destroyed between 1972 and 1999 (Gilpin, 2014). The institute began actively engaging in conservation, using Google mapping technology to collect data.

In 2006, the Institute launched its Gombe Chimpanzee Blog, providing findings discovered with daily updates from field researchers. This blog quickly evolved into a Google Earth layer which captures the work of the Jane Goodall Institute, including data about chimpanzees, along with their biographies. Included also was data on their habitats, interactive photographs, as well as video clips. All of these are overlaid on high resolution satellite images of the park, which illustrates the extent of the deforestation in the Gombe region. The Gombe Chimpanzee Blog is authored by field researcher Emily Wroblewski (Moore, 2007).

Historical Imagery

When opening Google Earth, the air photo and satellite imagery of the world that you see is relatively recent and updated on a regular basis. Even remote parts of the world, such as the Arctic Circle (on both the Canadian and Russian sides) include 2013 imagery. Thanks to the addition of the Historical Imagery time slider, users can now analyze and track the changes that occurred in many areas of the world over different time periods.

Some of the oldest imagery available is from the 1930s—aerial photos that were scanned and shared with Google Earth. Typically, however, photos are available for many areas of the world from the 1990s and onwards, and are usually in the form of satellite imagery.

The Historical Imagery feature is often used for personal use, as individuals enjoy going back in time to look at urban developments and familiar buildings to see how they have changed over time. However, there is tremendous value in exploring historical imagery in research as well. Many researchers have utilized the Historical Imagery time slider to measure changes in landscape and wetland development, urban growth, and population density fluctuations. For example, researchers looking at population increases and land use changes in the Rocky Mountain West region of the United States took advantage of the Google Earth Historical Imagery archive to compare one hundred wetland sites, going back ten years. The researchers then used the information to collect samples from at least 70 sites for their analysis (Johnson et al., 2012).

The Google Earth Historical Imagery resource has also been used to help investigate the looting of archaeological sites in Jordan. By identifying sites in Jordan and comparing them over time, researchers have been able to use Google Earth to quantify the damage and monitor them going forward (Contreras and Brodie, 2010).

Many people are familiar with the benefits of using historical images to capture drastic and sudden changes in landscape due to natural disasters. Google Earth is famous for being the first to share images captured right after a hurricane or tsunami swept through a city or state. Not only do the images show the extent of damage, but as people move on from these tragedies, the images remain as historical evidence and as a gage for how quickly or slowly a town was able to recover, evidenced by subsequent photos taken. In a way, Google Earth has become a preservation repository, archiving history through imagery. Peter Birch, Google Earth's product manager, stated, "A lot of this is about preserving our history and having a record of our footprint. It's easy to forget what the world looked like" (Peterson, 2009).

Whether it be natural disasters, human-made errors, or simply natural changes due to passage of time, Google Earth has been used as a platform for users to gain knowledge, exchange information, and receive updates. For example, when Hurricane Katrina struck southern Louisiana in August 2005, within two days Google Earth was used to open and view images taken by multiple organizations in the form of KML files and image overlays. By September 2005, Google and other organizations like the National Oceanic and Atmospheric Administration were regularly updating their satellite imagery allowing organizations to assess the conditions and the damage done (Alin, 2007).

Likewise, the Gulf Oil Spill in April 2010 had the entire world concerned about the people, environment, and the animals. Updates about how far the oil had spread were available via KML files, reflecting affected areas and allowing users to gain a better understanding of the gulf disaster (Seward, 2007). Visualizing the locations that are most in need makes getting help to them a much simpler task.

Google Street View

One of the most popular elements of Google Earth is its embedded "Street View" imagery. Launched in 2007, Google Street View provides street-level panoramic photography of urban and rural areas captured by special high speed, 360 degree cameras that are mounted on cars, bicycles, trolleys, and even personal backpacks. Street View gives the users the feeling of actually being on the streets, seeing the signs, traffic signals, houses, shops, bus stops, and people of that neighborhood. The images also capture the culture, as interpreted by the individual's clothing, in the modes of transportation, in the infrastructure of buildings, and in the types of commercial outlets, allowing users to conduct virtual audits as they move through the Street View imagery.

You can find the Street View imagery by selecting and dragging the "Pegman" character found in the Google Earth Navigation Tools over a street or group of streets in Google Earth. If the streets become highlighted as the "Pegman" moves over the map, then imagery can be accessed by dropping the character onto one of the highlighted streets.

Initially available for just a few American cities, Google Street View photography has rapidly grown in popularity with much of the United States and Canada photographed by 2009, and several European and Asian countries also signed on. By the beginning of 2014, Google Street View imagery was available in many of the world's major cities and rural areas, covering to some extent all seven continents. According to the 2014 Google Street View coverage map, approximately 35-40 percent of the Earth's surface included some

Street View photography. Interestingly large swaths of Asia and Africa still remained untouched, and only about half of South America had been captured in Google Street View.

Google Street View was an instant hit with the public when released, with many people checking their neighborhoods to see if they had been photographed yet. The visit of the "Google Street View car" to cities and towns in the United States and Canada often made local news headlines, with residents discussing where and when they witnessed the Google vehicle. It also did not take too long for Google to realize that they were not able to capture every location with their vehicle and started capturing 360 degree images using bicycles. To further expand its ability to take imagery, Google began a trekking program to encourage organizations to hire people to walk in obscure locations or even inside of buildings using "trekkers"—backpacks equipped with borrowed Google cameras. In one example, two men in Florida were paid twenty-seven dollars per mile by a tourism organization to take photos of hundreds of miles of beaches (Circa, 2014).

Trekkers have been used to capture the interior of buildings for a couple of years now, photographing businesses, tourist buildings like skyscrapers, art galleries, museums, and most recently even hotels. Users can simply click on the building of interest and they will be swept through to the interior, wrapped in 360 degrees of interior imagery.

With Google expanding its services down the "beaten path" as it were, many landforms and tourist attractions continue to be photographed and virtually visited by users. Some popular examples include the Grand Canyon, allowing users to get a close look at the river and the view from the bottom. The NASA Kennedy Space Center is another popular site, including images of some of the main areas like the launch pads.

Moving away from land, Google Street View has also expanded into the ocean, photographing marine life in their natural habitat. Scuba Divers are decked out with the cameras as their experiences are photographed and shared with the world.

Uses of Street View

Having virtual access to many popular and remote areas of the world has proven to satisfy the curiosity of many Google Earth users. From hobbyists, to not-for-profit organizations, to elementary students, to entrepreneurs, Google Street View is being used in various project works, such as hyperlapse videos (<http://labs.teehanlax.com/project/hyperlapse>), video games (<http://geoguessr.com/>), (<http://www.wonder-tonic.com/zombie/>), and music videos.

While people found it fun to check out and view their homes and neighborhoods in Google Street View, it was not long before academic researchers grasped the potential of the Street View imagery as well. In fact, several research papers have been produced since 2010 which explore the uses of Google Street View as a neighborhood auditing instrument. Using cities such as Chicago and Boston, researchers compare the results of a virtual (Google Street View) audit versus the in-person walk through looking at such factors as quality of life, healthy food choices, and presence of crime. Moreover, researchers can go through and add *Placemarks* in the Street View Imagery. While researchers generally still prefer the in-person method of gathering finely detailed neighborhood data, Street View has been lauded as a time efficient and low-cost alternative tool for neighborhood research (Clarke et al., 2010; Rundle et al., 2011; Ben-Joseph et al. 2013).

One study presented a new GIS method for measuring distance and coverage of sidewalks using Google Earth's tools, including the use of Street View. The study was interested in calculating sidewalk length, a feature not always readily available as a GIS file.

The researchers visually examined road segments on Google Earth using a combination of aerial photography and Street View imagery. They were able to successfully identify the presence of sidewalks, and therefore the walkability of certain areas in the neighborhood (Janssen and Rosu, 2012).

Essentially any feature of the road and surrounding areas can now be studied thanks to the Street View. Researchers at the University of Maryland have taken advantage of these images at their fingertips and found a way to identify thousands of potholes, obstructed pathways, missing ramps, cracked streets, and other road problems. Once identified, the images are saved and incorporated into a report for the local council's road-mending crews to review (Marks, 2013).

Most recently, Google Street View has been making their older archived Street View images available through Google Maps, creating a time slider to allow users to compare changes at street level dating back to 2007. Historical Street View will be made available in fifty-five countries, allowing users to observe changes in buildings, on lots, and in landmarks. Especially useful for social and economic studies, changes in built-up areas reflect good and bad times, such as bankruptcies, rotations of retail outlets, foreclosures, or new developments. Perhaps the most powerful example of using Historical Street View is in research looking at the continued decline of inner-city neighborhoods in Detroit, particularly steepening since the 2008 recession (Feinberg, 2014).

Although very popular with the public, not everyone has been enthused about the availability of Street View imagery. Although Google makes efforts to blur some elements in the photos such as faces, license plate details, and other personal identifying details, privacy groups and various governments have still expressed reservations to Google's methods of capturing imagery and providing access to the wider public. For example, Google Street View imagery has been removed and discontinued in some places such as Greece and India. Google has also had to pay penalties for collecting public wifi data during the Street View image capture process, and also for violating individual countries' privacy laws. Google Street View is actually often used as an example of why privacy concerns and policies should be reconsidered by countries (Bradley and Clarke, 2011; Elwood and Leszczynski, 2011).

Google Earth Pro: Additional Features

While Google Earth is the more widely known application, researchers might also be interested in some of the extra functionality provided by Google Earth Pro. These functions include the ability to import spreadsheets and Esri shapefiles into Google Earth, the ability to geocode addresses, and the options to save and print the Google Earth imagery at a higher resolution. There are also additional measuring options as discussed above.

Many GIS users accustomed to working with products such as ArcGIS, have made the extra effort to obtain Google Earth Pro for its ability to bring in shapefiles. For example, James Taulman, an American naturalist studying animal distribution patterns using animal location data created contours using shapefiles in ArcGIS. These were then imported into Google Earth Pro in which he could show the data over the 3D Terrain model providing a vivid illustration of the patterns (Taulman, 2010).

Geocoding of address data is also another powerful feature of Google Earth Pro. For example, the California Highway Patrol (CHP) used Google Earth Pro as part of their efforts to record and then visualize fatal and severe injury car collisions over a ten-year period. Some of their data gathering and mapping involved creating ArcGIS files, but Google Earth Pro allowed for a batch geocoding process (Bigham et al., 2009).

Key Points

Google Earth is a powerful visualization tool that not only allows users to virtually fly to and visit nearly any location on the planet, but also provides a deeper understanding of our real world, including our people and our environment. This easy-to-use and interactive program has extensive capabilities, supporting external resources and thereby transforming the virtual globe into a mapping and research tool. With its high resolution imagery and so many geographic and environmental layers, users can connect and stay in touch with the world.

Google Earth relies on third-party organizations and individual contributions to keep the virtual world updated. With users sharing their talents, skills, and research, they can reach out to and influence so many more people. More and more organizations, including libraries, archives, and museums are sharing their resources using the Google Earth platform.

- Google Earth's interface is easy to work with
- Google Earth's general map making tools allow users to customize their maps
- The built-in Layers are a popular resource for research
- Historical imagery can be easily used for before and after analysis
- Google Street View has many personal and academic uses

The next chapter will showcase examples of how Google Earth is being used as a library catalog and digital repository for the purposes of describing, managing, and providing access to library collections and other resources.

References

Alin, Dustin. 2007. "Hurricane Katrina." Disaster Response. June 20. http://sites.duke.edu/tlge_mds43/hurricane-katrina.

Ben-Joseph, Eran, Jae Seung Lee, Ellen K. Cromley, Francine Laden, and Philip J. Troped. 2013. "Virtual and Actual: Relative Accuracy of on-Site and Web-Based Instruments in Auditing the Environment for Physical Activity." *Health & Place* 19: 138–150. http://dx.doi.org/10.1016/j.healthplace.2012.11.001.

Bigham, John, Thomas Rice, Swati Pande, et al. 2009. "Geocoding Police Collision Report Data from California: A Comprehensive Approach." *International Journal of Health Geographics* 8, no. 1: 72.

Bradley, Eliza S, and Keith C. Clarke. 2011. "Outdoor Webcams as Geospatial Sensor Networks: Challenges, Issues and Opportunities." *Cartography and Geographic Information Science* 38, no. 1. http://go.galegroup.com/ps/i.do?id=GALE|A254012697&v=2.1&u=utoronto_main&it=r&p=AONE&sw=w&asid=beae7c90f11c8072be3bdc6e85959a9f.

Cartography Associates. 2014. "David Rumsey Map Collection." *Cartography Associates.* June 5. http://www.davidrumsey.com.

Circa. 2014. "Two Men Walking Florida's Coastline with Street View Trekker Camera." *Circa News.* Accessed June 8. http://cir.ca/news/google-street-view-trekker-lending-program.

Clarke, Philippa, Jennifer Ailshire, Robert Melendez, et al. 2010. "Using Google Earth to Conduct a Neighborhood Audit: Reliability of a Virtual Audit Instrument." *Health & Place* 16, no. 8: 1224–1229.

Contreras, Daniel, and Neil Brodie. 2010. "The Utility of Publicly-Available Satellite Imagery for Investigating Looting of Archaeological Sites in Jordan." *Journal of Field Archaeology* 35, no.1: 101–114.

Elwood, Sarah, and Agnieszka Leszczynski. 2011. "Privacy, Reconsidered: New Representations, Data Practices, and the Geoweb." *Geoforum* 42, no. 1: 6–15.

Feinberg, Ashley. 2014. "Tracking Detroit's Decay Through Google Street View." *Gizmodo*, June 5. http://gizmodo.com/tracking-detroits-decay-through-google-street-view-1586631393.

Gilpin, Lyndsey. 2014. "How Jane Goodall Institute Uses Digital Mapping to Save Chimp Habitats in Africa, Empower Children in the US." *TechRepublic*. May 16. http://www.techrepublic.com/article/how-jane-goodall-institute-uses-digital-mapping-to-save-chimp-habitats-in-africa-empower-children-in-the-us/.

GrrrlScientist, 2013. "How Google Earth Helped Find Mozambique's Lost Forest of Mount Mabu." *SciLogs* (blog). December 9. http://www.scilogs.com/maniraptora/how-google-earth-helped-find-mozambiques-lost-forest-of-mount-mabu/.

Hong, Christine. 2013. "The Mirror of North Korean Human Rights." *Critical Asian Studies* 45, no. 4: 575.

Janssen, Ian, and Andrei Rosu. 2012. "Measuring Sidewalk Distances Using Google Earth." *BMC Medical Research Methodology*. Accessed June 20. http://www.biomedcentral.com/1471-2288/12/39.

Johnson, Pieter T. J., Jason T. Hoverman, Valerie J. McKenzie, et al. 2012. "Urbanization and Wetland Communities: Applying Metacommunity Theory to Understand the Local and Landscape Effects." *Journal of Applied Ecology* 50, no. 1: 34–42.

King, Rachel. 2012. "Google Maps New Covers 75% of Global Pop, 26 Million Miles of Directions." ZDNet. April 4. http://www.zdnet.com/article/google-maps-now-covers-75-of-global-pop-26-million-miles-of-directions/.

Kiremex. 2013. "How to Find Biking and Hiking Trails Near You." *An Introduction to WIKILOC.* http://www.saudicaves.com/mx/wikiloc/index.html.

Lee, Jacqueline. 2013. "A Survey of Transverse Drainages in the Susquehanna River Basin, Pennsylvania." *Geomorphology* 186: 50–67.

Marks, Paul. 2013. "Crowds Prowl Google Street View to Speed Road Repairs." *New Scientist.* February 9. http://www.newscientist.com/article/mg21729036.300-crowds-prowl-google-street-view-to-speed-road-repairs.html#.U5TpY01OUnM.

Moore, 2007. "Raising Global Awareness with Google Earth." *Imaging Notes Magazine* 22, no. 2 (Summer). http://www.imagingnotes.com/go/article_free.php?mp_id=97.

Peterson, Perry. 2009. "Google Earth Photo New York 9-12-1001." *Cave News* (blog). May 6. http://caveviews.blogs.com/cave_news/2009/05/page/10/.

Ruegar, Bruce F., and Emma N. Beck. 2012. "Benedict Arnold's March to Quebec in 1775: An Historical Characterization Using Google Earth." In *Google Earth and Virtual Visualizations in Geoscience Education and Research*, edited by Steven J. Whitmeyer. Boulder, CO: Geological Society of America.

Rundle, Andrew G., Michael D. M. Bader, Catherine A. Richards, et al. 2011. "Using Google Street View to Audit Neighborhood Environments." *American Journal of Preventive Medicine* 40, no 1: 94–100. http://dx.doi.org/10.1016/j.amepre.2010.09.034.

Seward, Miray. 2007. "Oil Spill in the Gulf." *Disaster Response*. June 20. http://sites.duke.edu/tlge_mds43/about/.

Taulman, James F. 2010. "Display of Animal Location Data and Kernel Home Range Contours in Google Earth Pro." *The American Midland Naturalist* 164, no 1: 157–164.

Tuttle, Benjamin, Sharolyn Anderson, and Russell Huff. 2008. "Virtual Globes: An Overview of their History, Uses, and Future Challenges. *Geography Compass* 2, no. 5: 1478–1505.

United States Holocaust Memorial Museum, 2009. Accessed June 8. http://www.ushmm.org/.
Wagner, Mary Jo. 2006. "The View from Google Earth." *Geospatial Solutions* (May): 23–26. http://www.nxtbook.com/nxtbooks/questex/gss0506/index.php?startid=23.
Wikipedia, 2014. "Google Earth." Last updated June 8. http://en.wikipedia.org/wiki/Google_Earth.
Williams, Caitlyn M., Gregory S. Baker, and Bradley A. Ault. 2012. "Enhancing Usability of near-Surface Geophysical Data in Archaeological Surveys via Google Earth." In *Google Earth and Virtual Visualizations in Geoscience Education and Research*, edited by Steven J. Whitmeyer. Boulder, CO: Geological Society of America.

CHAPTER 2

Discovering Library Resources with Google Earth

IN THIS CHAPTER

- ▷ Managing and discovering collections with pathfinders
- ▷ Offering direct online access to resources and collections
- ▷ Digitizing air photos and maps in KML format
- ▷ Collecting and using born digital geospatial data
- ▷ Working with government open data collections

MANY LIBRARIES ARE TAKING ADVANTAGE OF SPATIAL TOOLS to organize and deliver their collections and digital projects. Google Earth is an excellent online resource that can be used as a portal for pointing library users to their resources such as pathfinders and online collections. This chapter will highlight several initiatives that have piggy-backed on Google Earth's technology, with projects spanning from simple pathfinders that lead users to collections in print, to dynamic and interactive developments that offer access to digital collections with just a couple of clicks. This chapter will also explore how a range of academic, government, business, and community libraries and organizations have been able to create, collect, and manage spatial files for Google Earth visualizations.

Google Earth: A Resource Management System

As was discussed in chapter 1, Google Earth supports several file formats so library collections that are electronic or have been digitized and converted into file formats such as jpgs or KMLs can be viewed in Google Earth. Photographs, for example, can be imported into the program and can be dragged to and draped over the approximate geographic locations they represent. Also, tables with address information or GIS files like shapefiles can easily be converted into KML format for Google Earth viewing. Once converted into a supported format, libraries are well on their way to using Google Earth as a library catalog.

Accessing library resources online is easy and convenient when using an interface that users are familiar with. Over its ten years of public use, Google Earth has proven to be a popular tool for library users wanting to search resources by location, not necessarily by keyword. There are many ways to use Google Earth as a portal to the library. Some are using the application to view map or air photo indexes, or to locate books on a specific geographic location, while others are finding direct access to the collection itself through links to digital resources. Through a combination of using images, placemarks, links, overlays, and pop-up windows, library staff have discovered a popular way to reach out to library users.

Pathfinders to Resources and Collections

Either as a supplement to or a replacement of the traditional library catalog, some libraries, archives, and museums are using Google Earth as a management and discovery tool for their collections. Library materials which feature geographic locations such as places described in a book, in photographs, or in maps, can be linked in Google Earth and displayed in a library catalog offering users metadata about the items. These pathfinders send users to the physical library to retrieve maps and aerial photographs, and they offer the ability to search by geography and location of interest which is often difficult to do in a traditional library catalog which does not always have a geographic field recorded in the item's metadata or catalog.

Becoming more popular is offering users sample images of collections, allowing them to see the material before accessing the physical copies. Thumbnail images of collections like letters or photographs may be enough to engage the user to seek the resource. Google Earth is also known to offer supplementary material. Textbooks can now be purchased with accompanying KML files that send readers to locations that the text refers to. A fairly recent initiative, "Google Lit Trips" <http://www.googlelittrips.com/GoogleLit/Home.html>, offers teachers and students an interactive introduction to literature studies centralized around story locations. Users can read a book and then go to Google Earth to track story characters as they fly from one location to another, enabling them to better visualize the locations in the books.

But perhaps the most popular collections to manage and host on Google Earth are map resources. This is not surprising, of course, as Google Earth is a map itself, and many geographic collections require a map index as a finding aid to maps stored in cabinets. Unlike books and journals, many maps and air photos are not cataloged with unique call numbers. They often have their own numbering system and are typically searched for by using a print index. Topographic maps, geological maps, air photos, photo maps, fire insurance plans, and others use indexes to guide users to the library's collections. Although a time-consuming task, for years there has been no other way to discover these maps in the vast array of map cabinets. Today, many libraries have digitized the indexes using Google Earth's or Google Map's interactive mapping application to assist with navigation and orientation. Once users zoom into their area of interest, they are able to identify the corresponding paper map sheet numbers that depict that area. By having the indexes online, users can perform the research at home and then visit the library to acquire the physical maps. The next section will highlight a selection of libraries that have developed online indexes for their map and air photo collections and have overlaid them in Google Earth for public consumption.

Map and Air Photo Indexes

Index overlays in Google Earth are offered in a couple of different ways. Some of them are provided in a table or list format, sorted by type and year of map, as can be seen in figure 2.1, and others are offered seamlessly with a Google Earth plug-in built directly into the web browser.

With the list method, users can scroll down the list of available indexes and click on the one of interest which would then automatically open in Google Earth. Depending on the default settings of the computer, sometimes the user will need to save the KML file first to the local drive and then either double click on it to open in Google Earth, or open it from within the program.

With the map method, users would save a step by viewing the information directly on the website. The Google Earth plug-in inserts the virtual globe directly into the website so users don't need to use two programs to enjoy their resources. Once the information is loaded onto Google Earth, users can click on the information available, such as a flight line or a center point, to obtain the reference information needed to locate the actual product in the collection. Indexes not only provide reference information for the product, but in general they can relatively quickly tell users whether the collection covers their

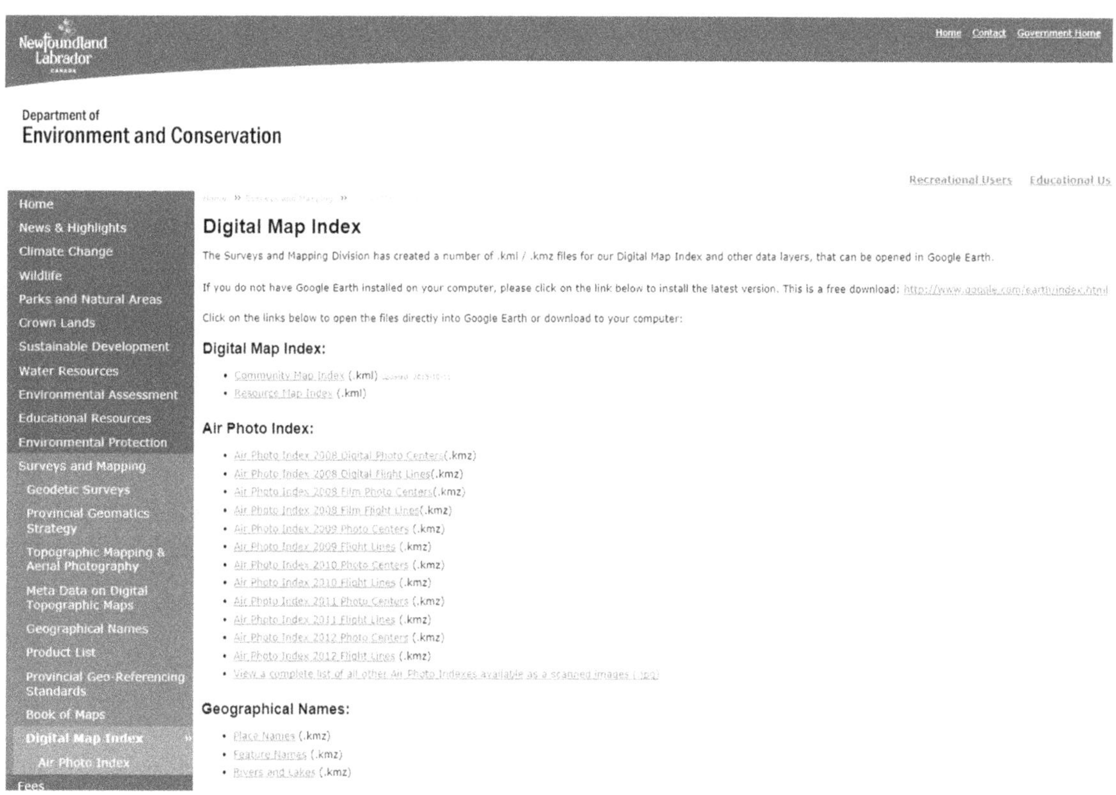

Figure 2.1. Digital Map Index

Source: Department of Environment and Conservation, Newfoundland Labrador: <http://www.env.gov.nl.ca/env/maps/digital_map.html>

actual area of interest. Although traditionally created to help aid in locating print material, more and more libraries are indexing geospatial data as well. Orthoimages, which tend to be offered in ten or twenty kilometer squares, or grids, require an index to organize and identify each unique image. Other datasets will use an index to display its geographic coverage. For example, the MacOdrum Library at Carleton University <http://www.library.carleton.ca/interactive-indexes> provides a list of available indexes for a variety of resources such as large scale topographic maps, geospatial vector data, orthoimagery, satellite imagery, census tracts, and more. Clicking on the census track index would display census boundaries and associated census track numbers.

Queen's University Library <http://library.queensu.ca/webdoc/maps/cdn_historic_topo_map>, similarly offers their map indexes online. Specializing in National Topographic Maps in a variety of scales, users can download the index and view it in Google Earth. These types of indexes can be easily created by converting the original shapefile into a KML file. Many indexes are originally created using a GIS program, so users will find files in shapefile format. Creating original KML files is possible as well as is converting from other formats and this will be discussed in chapter 4. Three different library indexing projects provide examples—McGill University Library's Air Photo Indexes, University of Toronto Mississauga Library's Air Photo Indexes, and the University of Connecticut.

McGill University Library's Air Photo Indexes

McGill University Library has a collection of nearly 50,000 air photos that have not been catalogued and require indexes to help locate them. Indexes have been created and made available for many of their air photo collections covering the city of Montreal, the province of Quebec, and various other regions within Canada. The website <https://www.mcgill.ca/library/find/maps/airphotos> offers KML files to view the indexes in Google Earth. Once loaded into the program, users can browse the index by zooming into the area of interest and clicking on the points, each representing the center of the photograph. Once clicked, a balloon pops up with the aerial photograph information, including all the elements that are needed to retrieve the photo in the print collection—photo year, flight line number, photo number, and scale. Users can also systematically browse through the folders available under Places in Google Earth. There is a folder available for each unique year (i.e., 1928, 1993), and when users click on the folder, it further separates the flight line numbers (i.e., A26273). Clicking on the flight line number folder will lastly provide the photo number (i.e., 089). Clicking on the photo number will zoom into the geographic area it covers. This index was created in the GIS software program, ArcGIS, and then it was converted into KML. These types of indexes are often created by populating a spreadsheet with the air photo details. Every point seen on the map is represented by a row on the spreadsheet, including the center of the air photo, denoted by latitude and longitude coordinates.

University of Toronto Mississauga Library's Air Photo Indexes

The University of Toronto Mississauga Library has a small collection of aerial photographs that cover portions of Southern Ontario for the 1950s. The indexes for these collections have been made available directly on the library's website by using the Google Earth plug-in. As long as users viewing the website have the plug-in installed, they will have a small version of Google Earth embedded into their Internet browser. The plug-in is free to install, but if users prefer, they can simply use Google Chrome as it already has the Google Earth plug-in available by default.

When visiting the index website <http://library.utm.utoronto.ca/datagis/airphotos>, the virtual globe is seen embedded into the browser, along with the index overlaid on top. Users can zoom into their area of interest to see if it is covered by the index, and if it is, zooming into it further will reveal flight lines color coded to represent different years. Each line provides precise air photo information including flight number and photo numbers. Once this information is obtained, users can visit the library in person to locate the air photos.

University of Connecticut Library Air Photo Indexes

The University of Connecticut Library, in collaboration with the Connecticut Department of Environmental Protection (CT DEP), has developed a collection of digital historical air photo indexes <http://magic.lib.uconn.edu/mash_up/aerial_index.html> for aerial photography of the state of Connecticut dating from 1934-2010. The individual indexes can be viewed in Google Earth, or by using GIS software. A Map Preview link allows the user to see the geographic extent of the photography available before downloading the index itself.

When the user clicks on the KML link, Google Earth will automatically load, displaying the file. The index appears as a series of points, with each point representing the center of the air photo. When the user clicks on the point, the air photos' metadata pops up including photo information like year and scale, and a preview of the image itself. The University of Connecticut Library has digitized over 20 air photo indexes.

Other Collection Holdings

Although it is common to use maps as interfaces to locate cartographic material, most recently libraries, archives, and museums have started using Google Earth as pathfinders for other collection holdings. Any collection that has significant ties with a particular place is an excellent candidate for being placed on the map in Google Earth. Whether they are pathfinders to books, journals, artefacts, or interviews, it is fairly simple to tag the collection record in Google Earth and offer a link to the library or museum's collection.

Bernice Pauahi Bishop Museum's Collection

One online project definitely worth exploring is at the Bernice Pauahi Bishop Museum, in Honolulu, Hawaii <http://www.bishopmuseum.org/special/googleearth.html>. The Bishop Museum offers exhibits and collections on the history, arts, and culture of the Hawaiian people, and is recognized throughout the world for its scientific research and educational programs.

The Bishop Museum's website offers online exhibits including two KML projects that consist of thousands of museum records. The Collections of Mauna Kea Natural Sciences, Library and Archives holdings include the museum's entomology, botany, vertebrate, and invertebrate zoology collections, as well as the Invasive Marine Species of Hawaii project, a collaboration between the Bishop Museum and the State of Hawaii Division of Aquatic Resources, consisting of nearly 200 collection locations of invasive marine species.

The Mauna Kea Collections are comprised of the Natural Sciences Collections (mammals, entomology, birds, and botany) and the Library Collections. When clicking on any of the placemarks in Google Earth, a balloon pops up with a number of holdings related to that geographic area, and a direct link to those library holdings on the Bishop Museum's website. The holdings include author, catalog number, collection title, keywords, and publication date. Since many museums don't offer their collection records online, using Google Earth as a content management system not only promotes the museum's collections but does so in an innovative way.

Although some of these projects look fairly sophisticated and it may appear that they are difficult to create, it is in fact quite simple to create a collection catalog in Google Earth. A spreadsheet (like Excel) would need to be created and populated with the information that you would like the users to see when they click on the placemark. Information about the collection and links to an online catalog or page describing the holdings can be added. Photographs can be included, and links to videos are possible. One mandatory column the spreadsheet requires is the locational information—latitude and longitude of each geographic location the record relates to. The spreadsheet can then be added into Google Earth, and placemarks are automatically created, displaying one placemark for each row of information.

Portland State University Library's Yizkor Book Collection

The Portland State University Library's project is a model for all libraries as it showcases how book titles can be managed and promoted in Google Earth. The Portland State University Library has a significant collection of rare publications from the Middle East, including over 130 Yizkor books.

Memories of the Jewish communities and families through photographs and historical narratives are available to be viewed via Google Earth <https://library.pdx.edu/yizkor.html>. Users can download the KML file available from the website and explore the compilation of resources pulled together in the format of placemarks denoted by Stars of David.

Each placemark represents the location of a community, family, or individual affected by the Holocaust. Each record includes a link to the corresponding Yizkor book record, either available at Portland State University or the New York Public Library. Some records also include links to historical Jewish museums, genealogical societies, memorial websites and the book's table of contents. This is an excellent project that encourages collaboration with other libraries and museums.

YIZKOR BOOKS

Yizkor books were written after the Holocaust as memorials to Jewish communities devastated by the Holocaust. They were typically written by survivors from those communities and contain descriptions and histories of the community, biographies, and lists of people who perished.

GOOGLE TOUR BUILDER

Google Tour Builder is an online storytelling tool that integrates place and the story's text, photos, and videos. The application creates tours that fly from one place to another along the storyline of your map.

Harvard Law School Library's Historical Recent Acquisitions

The Historical and Special Collections Department in the Harvard Law School Library is comprised of a treasure trove of rich materials spanning ten centuries. With so many items being too rare and fragile to touch, the library has offered a digital window to get a glimpse of the books and manuscripts. Comprised of photographs of their collections, the library has created an online exhibit called *Spanning the Centuries: An Exhibit of Recent Acquisitions, 1579-1868*. The exhibit was also shared via Google Earth's new Tour Builder application <https://tourbuilder.withgoogle.com/builder#play/ahJzfmd3ZWItdG91cm-J1aWxkZXJyEQsSBFRvdXIYgICAsNmgtAsM>, a wonderful new tool that puts the items in the exhibit in their geographic context.

Google Tour Builder is embedded into the web browser so users will need to either use Google Chrome, or they will need to download the Google Earth plug-in for their browser. The Harvard Law School Library's historical collection is captured through photographs which are tagged to specific geographical locations across Europe and North America. The online exhibit includes the Google Earth map on the right side of the browser, and a slide show with photos and descriptions on the left. The slideshow can be played automatically so as each slide advances, Google Earth spins and zooms into the next corresponding geographical location, all while displaying and forwarding through the digital exhibit on the left.

Direct Resource Access

Although pathfinders, such as indexes and resource links, serve to bring users to collections, they still require their users to make in-person visits to the physical repository. Even though the experience of leafing through original documents cannot be replaced with their digital counterparts, there is something to be said about having seamless access to digital collections. Many documents, maps, photographs, even books have recently become digitized, and many libraries are creating websites to host and manage their digital repositories. Some libraries are embedding Google Earth into their Internet browsers, using the virtual globe to display the actual collections. Others are offering their online resources in KML file, downloadable and viewable on local copies of Google Earth.

The benefits of using Google Earth to organize and display collections is certainly seen in cartographic products, but more recently seen in collections like videos, audio recordings, photographs, and documents. Again, any material with a geographic association can be mapped in Google Earth, with its free availability and ease of use making it a popular choice for organizations such as libraries.

Map and Air Photo Collections

Very popular with map and air photo collectors, Google Earth does a fabulous job georeferencing the product to ensure near precise placement over the virtual earth. Maps and air photos are often stored and displayed in Google Earth, offering users interactivity features like zooming and overlaying other data like roads and geographic boundaries. In lieu of using a photo hosting application, Google Earth not only displays the digital collection, but also helps describe the collection with the various features embedded within the program.

Libraries and historical societies take special advantage of the Google Earth image overlay, offering the public direct access to their scanned and digitized maps. The maps are often historical in nature, both due to the large interest in historical mapping but also because of copyright restrictions on modern maps. A simple way to offer map overlays in KML format is to list the files on the organization's website. Many projects include a map thumbnail which, when clicked, opens the map automatically in Google Earth. The KML files themselves are quite easy to create. Maps can be georeferenced in a GIS program and saved as KML, or they can be brought into Google Earth and georeferenced directly there. Google Earth supports a wide variety of image formats.

One of the first collectors to adopt KML as a format for delivering maps was David Rumsey, a longtime map collector and philanthropist. Rumsey began collecting his personal map collection over twenty years ago and soon after began making scans of his vast collection available for free download from his website <http://www.davidrumsey.com>. Interested in maps, technology, and sharing his collection with the public, Rumsey continued to develop his website wanting his users to experience his map collection in new and interesting ways. Along with offering his map scans in different file formats and making them viewable in different browsers, Rumsey even included additional GIS layers and KML overlays to provide the user with a look at the past to see changes that had occurred over time. Passionate about sharing his maps with the world, Rumsey's partial collection became a permanent addition in Google Earth. The Rumsey collection includes maps from around the world with a strong emphasis on the American west.

Rumsey's innovative online collection has inspired many others to do the same. Today, there are many digital online projects that offer maps and other resources like aerial photography for viewing in Google Earth. One example of accessing aerial photography can be seen through British Columbia's GeoBC website <http://geobc.gov.bc.ca/base-mapping/imagery/airphoto-viewer.html>. Users can simply download the KML file and view the flight information as well as the low resolution photos. A typical map overlay project can be seen at the Minnesota Historical Society website <http://collections.mnhs.org/maps/index.php/10002223>. The Minnesota Historical Society's collection includes about 19,000 maps reflecting the state of Minnesota from 1849 to the present. A small collection of these maps are available for viewing via Google Earth.

Likewise, the Lakeshore Genealogical Society <http://www.lakeshoregenealogicalsociety.ca/history-area/maps/northumberland1878> is compiling and sharing historical cartographic collections. A collaboration with Rare Books and Special Collections, McGill University Library, the Lakeshore Genealogical Society is offering the Canadian County Atlas Digital Project. Published between 1874 and 1881, the Canadian atlases provide township and property information. Northumberland County is available as

Google Earth map overlays and can be accessed by downloading the KMZ file. Included in the file are county and township maps as well as photographs.

There are some online projects that offer a bit more than a simple KML file to download. Some have embedded Google Earth into their Internet browser, and others offer an exhaustive online collection. Three projects worth highlighting include Georgia State University Library's Digital Collections, the University of North Carolina's Historic Overlay Map Collection, and the University of Waterloo Library's Historical Map and Air Photo Collections.

Georgia State University Library's Digital Collections

Georgia State University Library has an extensive digital collection encompassing a wide variety of themes, from maps to oral history projects to photographs. One of the most popular collections include The City Planning Maps <http://digitalcollections.library.gsu.edu/cdm/search/collection/atlmaps>, with over 1,500 maps from the City of Atlanta, Atlanta Regional Commission, and other agencies. These maps tell the story of Atlanta's development from population density, to land use patterns, to building projects and infrastructure development. The collection is freely available to the public and can be viewed in Google Earth. The collection, represented by thumbnail images, can be browsed and once selected can be viewed in the browser. There are several format options available for downloading, including Google Maps, Google Earth, GeoTiff, and jpg. Once opened in Google Earth, the map displays over the corresponding air photo and users can use the transparency tool slider to compare the before and after to analyze changes in the city.

The Planning Atlanta Collection also includes 124 individual aerial photographs representing the highly populated city in 1949. The photographs were mosaicked to create a seamless overlay, and both the mosaic and the individual photographs are made available for viewing and for download from their website <http://digitalcollections.library.gsu.edu/cdm/search/collection/atlaerial>. Each image is represented by a thumbnail and can be downloaded for Google Earth viewing. The Google Earth mosaic, created from the 124 original photos, is a large file (over 1.6 gigs) that covers the City of Atlanta.

University of North Carolina Library's Historic Overlay Map Collection

North Carolina Maps <http://www.lib.unc.edu/dc/ncmaps/> is a historical map digitization project that is a collaborative effort by the North Carolina State Archives, the Outer Banks History Center, and the University Library at UNC-Chapel Hill. The project is comprised of pre-1923 maps and atlases covering the regions of the state of North Carolina, and beyond. The maps that have been selected to be viewed in Google Earth are available to be browsed <http://www2.lib.unc.edu/dc/ncmaps/interactive/overlay.html>. What makes this project a little different from some of the others previously discussed is that Google Maps has been embedded directly into the website so users are able to enjoy viewing the maps seamlessly. If, however, users would still like to download the KML file for viewing in Google Earth, they can do that as well.

When a particular map link is clicked, a webpage opens with Google Maps in the center and the map controls on the left side. Users can turn the historical map on or off, as well as zoom and activate the transparency feature. This map display can be seen in figure 2.2.

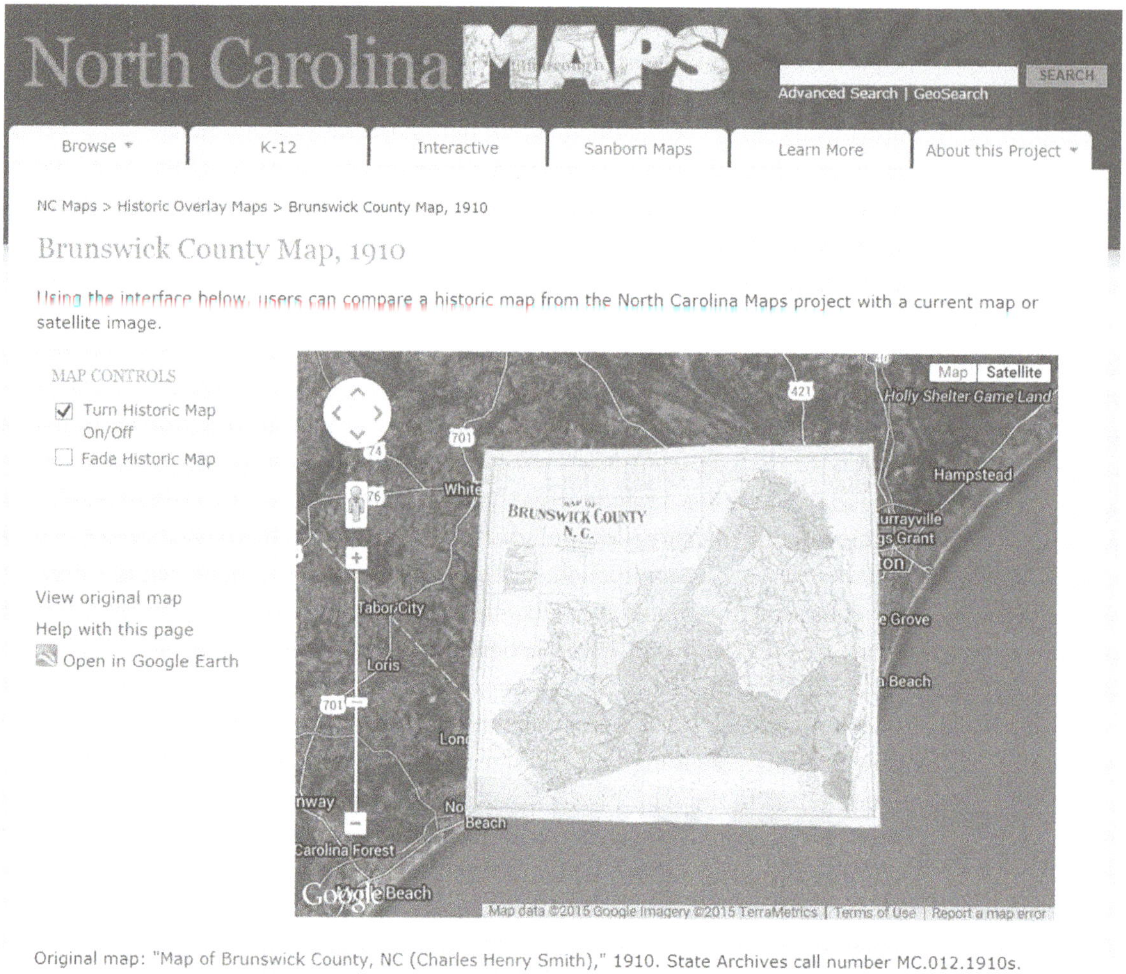

Figure 2.2. North Carolina Maps: Brunswick County Map, 1910

Source: North Carolina University Library: <http://www2.lib.unc.edu/dc/ncmaps/interactive/MC_012_1910s.html>; Google Maps https://www.google.com/ Map data: 2014 Google, 2014 TerraMetrics

University of Waterloo Library's Historical Map and Air Photo Collections

The University of Waterloo Library has created a couple of online projects that deliver their specialized collections using Google Earth technology. Historical maps and air photos have been scanned, georeferenced, and shared online for users to freely view and download. One initiative is the Waterloo Region Historical Map project <http://www.lib.uwaterloo.ca/locations/umd/digital/historicmaps.html>, offering the public a small collection of historical maps for the Region of Waterloo. The maps can be viewed directly in Google Earth, which is embedded into the website (plug-in required). Users can select a map of their choice, and it will appear overlaid in the correct location on Google Earth. Reference details about the map will also appear along with a link to the item's catalog entry. The maps represent the history of the region between 1815 and 1923.

Another project created by the University of Waterloo Library is the Historical Air Photo project <http://www.lib.uwaterloo.ca/locations/umd/project/>. This initiative involved scanning approximately 2,000 historical air photos for years between 1930 and 1963 for the Kitchener-Waterloo area. The website offers a map interface to zoom in and select the area of interest. Once the geographic location has been identified, users are provided with a table to download the actual air photos for varying years in a number of different formats, including KML. Users can click on the KML file and it will open in Google Earth, displaying the air photo in the correct location, allowing users to then use the transparency tool to compare the historical air photo to the modern imagery Google Earth supplies.

These types of digitization projects have paved a new path for libraries and their users, as not only are the resources becoming easily accessible online but they are being referenced to the Earth's surface, neatly overlaid to allow users to study, identify, compare, and make conclusions about the landscape's past and present. With the growing popularity of KML files, the general public is now becoming more comfortable using them, allowing libraries to concentrate on a standard geospatial file format for disseminating digital collections to non-traditional map users.

"Born Digital" Geospatial Library Collections

Up until this point in the chapter, the focus has been on how libraries have used Google Earth as a pathfinding tool for existing paper map, aerial photo, and other library collections. It was also demonstrated how some libraries have taken the next step in scanning their print resources into Google Earth KML files. Without a doubt, Google Earth is stirring up notable changes not only in library communities, but throughout many other groups and organizations that use geographic information to communicate research, activity reports, and other data. In fact, Google Earth allows users to both see and create spatial data, literally opening up new communities of users to the powers of "born digital" geographic information.

Interestingly, Steven Morris, a librarian at North Carolina State University anticipated the dramatic evolution that Google Earth was going to unleash, publishing in *Library Trends* that Google Earth was "likely to be just the beginning of a revolution in how geospatial content is used and republished." Morris would go on to note that:

> The geospatial content available in these environments is still limited, with only a very, very small slice of all available geospatial content exposed for use . . . however, holes are being poked and then widened through the walls . . . (through the conversion of) . . . traditional geospatial data to Keyhole Markup language (KML) for integration with Google Earth. (Morris, 2006, 295)

Excited about the new opportunities that Google Earth will open up for libraries, Morris declared that: "One very immediate impact of Google Maps, Google Earth, and the like will be the creation of a much larger audience and market for geospatial information resources" (Morris, 2006, 295). The larger audience and market for Google Map and Google Earth predicted by Morris very quickly developed the KML file, adopted by technologists and application developers as a standard in which to develop the "Geoweb." In 2008, the popularity of Google Earth was formally recognized when the Open Geospatial Consortium (OGC) approved the Keyhole Markup Language (KML) as an "open standard," noting that the format allowed both novice and expert users to publish and share geographical information in an open format. One observer even noted that KML was doing for geographic information what HTML did for web content, allowing the free flow of information in an online environment (Bacharach, 2008).

With the OGC recognition of KML as an "open" format, libraries were quick to adopt Google Earth as a means to visualize their digital collection projects, particularly if the project had place-based information that could be illustrated in a digital map. As discussed earlier in this chapter, Google Earth has been used to provide finding aids for library users seeking aerial photography or special map collections. Some libraries have taken the adoption of Google Earth one step further by actually creating KML collections of digital material and providing them for their users. Duke University Libraries, for example, has taken their historical geospatial files of the Duke Forest Lands and made them available for research and teaching <http://library.duke.edu/data/collections/gis/forest>. Available for the time between the 1930s and 1980s, these layers include Cultural Operations, Disturbances, Harvestings, Plantations, and Standings.

In fact, several special collections around the United States have utilized Google Earth in interesting ways. One approach can be found at the Belgian-American Research Collection at the University of Wisconsin <http://digital.library.wisc.edu/1711.dl/WI.BelgAmrCol>. With Google Earth, users of the collection can tie distinctive survey maps of Belgian settler farms to their property locations in the Wisconsin countryside. The maps overlaid in Google Earth are also linked to the photo records and thumbnail images stored in the Wisconsin databases providing users with multiple perspectives and access points into this unique collection.

Yale University Library has also found new ways to use Google Earth to maximize access to its popular Sanborn Fire Insurance Map collection. Yale staff have converted the Fire Insurance plans to a digital format and georeferenced them over the present day images of New Haven and other Connecticut towns. A 3D map of 1901 New Haven was even generated in Google Earth (see figure 2.3).

A similar project was also done in Galveston Texas where a library took its special *Galveston Storm Collection of 1900* and developed some KML files to highlight areas of flooding and damage and link it to the storm-related collections held at the Galveston

Figure 2.3. 3D Map of New Haven using 1901 Sanborn Fire Insurance Plans

Source: Yale University Map Collection, <http://www.library.yale.edu/MapColl/print_sanborn.html>. And Google Earth 7.1. Map data: 2014 Google

and Texas History Center <http://www.rosenberg-library.org/collections/1900storm/google.earth/google-storm.html>. Texas A & M University also has created a KML index for its scans of the early twentieth-century Geologic Atlas of the United States <http://repository.tamu.edu/handle/1969.1/2490>.

Outside of special projects, however, many institutions have recently begun offering their regular geospatial collections in KML format. As one of the early innovators in providing GIS datasets and services in academic libraries, it is no surprise to see the "MAGIC" at the University of Connecticut (UConn) Libraries take a lead in offering geospatial data in KML <http://magic.lib.uconn.edu/connecticut_data.html>. The UConn MAGIC staff have made all of their Connecticut data available as KML files including boundaries, environmental layers, roads, and transportation files among others. Even more striking about their collection is that they both have current and historical layers and they are not just restricted to UConn students, but open to anyone who wishes to download the files.

Many libraries have adopted the University of Connecticut Library's approach to disseminating their own data using more than one geospatial format. Libraries typically offer geospatial data files in Esri shapefile formats, but more libraries are now converting these files into KML to attract a broader user base. Geospatial data has traditionally been offered to users locally, offline and in person. Many libraries still offer their datasets in this way mainly due to the infrastructure costs of storing data in the cloud, but over the last several years we have seen a considerable increase in online geospatial portals. Libraries are differentiating between traditional GIS files and KML files by often categorizing the Google Earth files as "Overlays." More and more libraries offer an Overlays section on their website that is dedicated to KML files.

Columbia University in the City of New York has a *Columbia University Spatial Data Catalog* <http://gis.columbia.edu/data.html>, enabling users to search specifically for datasets in KML format. In the search field, simply type in "KML" and a listing of all corresponding files will appear. Users can then access the metadata and the file itself; however, due to licensing restrictions, only those affiliated with the institution can download the resource.

The University of California at Berkeley developed a geospatial data repository, *GeoData@UCBerkeley* <http://gis.lib.berkeley.edu:9081/> that contains over 9,000 geospatial data layers from UC Berkeley, Harvard University, and MIT. The repository is part of a consortium of geoportals created by Tufts University, Harvard University, MIT, Princeton University, and the Massachusetts Office of Geographic Information (MASSGIS). Using an in-house mapping application called MapUp, users can not only search and access data, but being a repository, they can also upload their own datasets. Users can search the database by topic, keyword, data originator, year, and data type (point, line, polygon, raster), then zoom into their area of interest and preview the layers before adding them to the "cart" for free downloading as either a shapefile or KMZ. GeoData is an excellent example of innovative collaboration, as well as a model for future cataloging and data management projects. Collections that have a geographic component should be searched spatially, and more and more geoportals are offering this type of online platform.

A major undertaking by academic libraries in the Canadian province of Ontario is the collaboration and development of the Scholars GeoPortal <http://geo2.scholarsportal.info/>, a geospatial data-discovering and analysis tool created specifically for university students and researchers. The datasets are abundant, including land-based vector data like water and culture features, transportation networks, geology, soils, census data, orthoimagery, and more. The portal offers search, preview, query, and download functionality (including KML format) for datasets licensed by Ontario university libraries.

With all the effort that libraries are putting into offering collections and geospatial data in Google Earth format, they subsequently need to promote and teach library users and library staff how to use the mapping application. Even though many students and researchers are fairly familiar with the program, there are even more that are not and can benefit from taking a workshop, viewing a webcast, or going through some tutorials.

Google Earth resources are often promoted on library websites and subject guides. Many libraries create their own customized instructional guides including videos, some also offer Google Earth training sessions and workshops. The University of Connecticut, for example, offers a resource page <http://magic.lib.uconn.edu/help.html> that includes four different instructional videos covering geotagging photos, historical imagery, importing KML and GPS data, and recording tours. Likewise, Thompson Rivers University in British Columbia offers Google Earth resources in a subject guide format <http://libguides.tru.ca/content.php?pid=250564&sid=2299569>, displaying Google Earth videos, RSS feeds, and tidbits on Google Earth features. Wesfield State University also has a devoted subject guide to Google Earth <http://lib.westfield.ma.edu/content.php?pid=588203&sid=4865287>, offering users a one-stop shop for information on the application, tutorials, and articles on KML files and links to free KMLs. Also very helpful is information on how to obtain Google Earth permission to use their resources in publications.

Perhaps the most popular resource found on library websites are customized tutorials offering users specific instructions on how to use some of the features available in the program. The University of Waterloo, for example, has six different Google Earth/KML tutorials teaching users anything from how to create walking tours to how to convert shapefile to KML. Carleton University also offers an abundance of resources teaching users the basics of getting started with the program to enabling developers to embed Google Earth applications into webpages using Google Earth Application Programming (API).

Libraries are also taking it a step further and offering these tutorials and user guides live, in the form of workshops. Workshops can range in duration from one hour to a full day and are typically offered in the library as part of a GIS teaching program. Once users are trained on the basics of Google Earth, they are able to delve into more advanced capabilities of the application by taking other workshops or by learning on their own using the vast amount of resources available online.

Although libraries have traditionally been the gatekeepers of geospatial data, especially licensed data, more and more users are discovering online data resources outside of the library. Many organizations, such as governments, are storing their own information in KML formats, and making them available online. Opening up and providing their information to the public contributes to this bank of resources that users are hungry for. Instead of collecting only commercial data resources, libraries can now point users to other products. It is quite common for libraries to manage external links on their subject guides and other webpages. When a new external resource is discovered, the library adds it to the list, providing its users with a comprehensive collection of internal and external resources. The following will examine external data more closely, looking at open government resources—free open data that anybody can access.

Government Open Data Collections

The term *open data* refers to the belief that datasets collected by government agencies on behalf of taxpayers should be made available to all citizens without restrictions (Wikipedia, 2014). Although the principles underlying the concept of open data have been present for many decades, especially in the United States where government information has long been in the public domain, it has only been in recent years that geo technology such as Google Earth has taken the awareness and usability of government open data to a whole new level of public interest.

Prior to Google Earth's launch as a free and easy-to-use mapping application, there were very few publicly available tools to view, let alone manipulate, government-derived geospatial data. Only users familiar with specialized Geographic Information Systems (GIS) software could utilize such data. As such, governments were under very little public pressure to share their geospatial data, at least in a way that could be reused by the masses. Some governments, particularly in the United States, nevertheless still shared their datasets as part of a government mandate for sharing public information, but did so knowing that only GIS specialists in other government agencies or academia would ever be interested in the data. In their eyes, the potential for reuse of the data was very limited.

With the arrival of Google Earth for public download in 2005, the interest and demand for geographic information changed almost overnight. For the first time, the public became interested in looking at not just Google Earth imagery, but also what else might be out there. Accessing information about the world now seemed easy and open. For the first time, anyone with a computer and Internet connection could easily view imagery of their homes, cottages, famous cities, and landmarks all over the world. Moreover, they could also add their own data layers and mark up the imagery to create their own maps or KML and KMZ files. Within weeks of Google Earth's release, people began sharing their own user-created KML files, and some government data providers began making datasets available in KML format.

United States

A belief in open data in the United States is clearly evident when visiting one of the federal government websites, particularly those involving science and research. The United States Department of Agriculture (USDA) Forest Service website includes a page devoted to "Fire Data in Google Earth" <http://activefiremaps.fs.fed.us/googleearth.php>. The site includes a plethora of current and historical data including KMZ files covering Fire Detections, Incidents, and Fire Potential.

The United States Geological Survey (USGS) Earthquakes program is another site providing KML files for users interested in mapping earthquake data <http://earthquake.usgs.gov/learn/kml.php>. Even more impressive is that the data are available in real time. The Federal Emergency Management Agency (FEMA) also has flood data available <https://hazards.fema.gov/femaportal/wps/portal/NFHLWMSkmzdownload>; and the National Aeronautics and Space Administration (NASA) provides KML data, particularly from the Jet Propulsion Laboratory (JPL). On the JPL website, a selection of mosaic views of the Earth that can be overlaid in Google Earth are available <http://onearth.jpl.nasa.gov/KML.html>. NASA also has amazing visualizations from Goddard Space Flight Center that are available for viewing in Google Earth <http://svs.gsfc.nasa.gov/documents/available-Biosphere.html>.Perhaps the most fascinating among the NASA KML collections is found at their website dedicated to discoveries made by the Hubble Space Telescope. Rather than having KML data focused on Earth-bound features, the Hubblesite.org has KML files of the telescope's location around the Earth. You can also switch to "Google Sky" mode to view Hubble data on various comets and supernovas as they happen far away from Earth <http://hubblesite.org/explore_astronomy/gsky/kml_collection.php>.

Finally, the National Oceanic and Atmospheric Administration (NOAA) provides a treasure trove of current and historical weather data from their National Weather Service webpages in KML/KMZ format for public download <http://www.srh.noaa.gov/gis/kml/>. Of course, it is not only science-related topics that have benefited from the release of Google Earth. The United States loves to promote and celebrate its history, so not surprisingly, the National Register of Historic Places has put up a KML file documenting its locations and respective histories <http://nrhp.focus.nps.gov/natreg/docs/Download.html>. Even the National Museum of Art has climbed aboard the Google Earth bandwagon with its Afghanistan treasures exhibit now featured in a Google Earth layer <http://www.nga.gov/exhibitions/2008/afghanistan/burialsites.shtm>.

Perhaps the most significant indication of the synergy between open data and the availability of Google Earth/KML files in the United States came with the 2009 launch of the Data.gov website <https://www.data.gov/>. Not only is U.S. federal government data available from the portal, but various state and local government datasets can also be found by searching through this portal. For example, just doing a search for "KML" brought up over 2,700 results including data as varied and diverse as on NOAA's Ship Tracker and 10 foot Contour intervals for the City of Baltimore.

Moving from the federal to the state level of government, you will also find a healthy open data culture, which has actually been functioning, if not thriving, for many years before the arrival of Google Earth. Although the term open data had yet to be coined, many state governments by early 2001 had already established their own geospatial "clearinghouse," where state-produced data was served up to be shared among state agencies and taxpayers alike usually in Esri shapefile or CAD format. The release of Google Earth simply served to make state-produced data even more accessible, and within a couple of years, most states followed up by making KML file formats also available for each of their datasets.

The Commonwealth of Pennsylvania, for example, manages PASDA, which is the Pennsylvania Spatial Data Access website <http://www.pasda.psu.edu/>. This serves as the Geospatial Data Clearinghouse for all state-produced geospatial data. Best of all, almost all files are provided with a link offering the user to "View in Google Earth." Along with having many files available in KML format, this site is noteworthy for also containing historical GIS files, such as Ecosystem analysis of urban forest from the present and going back to the mid-1980s.

Another example of a state geospatial data clearinghouse website is the "Maine Office of GIS Data" <http://www.maine.gov/megis/>, which includes a data catalog, online mapping, and showcase for mapping projects from a Google Earth perspective. The "Maine Office" is especially noteworthy as it provides a direct link to the file in a KML format, rather than having to check to see if the download options include KML.

Moving down from state to local (county and municipal) governments in the United States, the availability of Google Earth has probably brought the biggest change in open data access. Unlike at federal and state levels, many local governments were reluctant to share their data prior to the launch of Google Earth in 2005. Most local governments were content with either keeping them in-house or selling them to users with strict licensing policies. For example, the Metro Regional government of Portland Oregon sold "Metro GIS Datasets" on CD-ROM to anyone interested for many years beginning in the late 1990s. Today, however, Portland and many other cities are now making much of their data available as "open," while still supporting an online subscription service for some datasets <http://rlisdiscovery.oregonmetro.gov/>.

Although never mentioned explicitly, it is interesting to note that many open data initiatives got underway in the months and years immediately following Google Earth's public release. In his excellent *Open Government Data: The Book*, Joshua Tauberer notes how Chicago was one of the first cities to see its crime data released as open data to be then published in Google Maps (Tauberer, 2014).

In looking at the City of Chicago Data Portal <https://data.cityofchicago.org/> you can quickly see why open data advocates use Chicago as a model for open data delivery. Data layers include ward and police boundaries, crime, business licenses, energy usage, health services, historic preservation, and much more. The sheer volume of Chicago-produced data available is impressive, and the data search interface web is very easy to navigate. Moreover, geographic-related data can be visualized by overlaying the data over a choice of Google Maps, Bing, or Esri Basemaps. The data layers can then be downloaded into a variety of formats including both Google Earth formats (KML & KMZ) among others.

Interestingly, the New York City Open Data Portal <https://data.cityofnewyork.us/> is similar to the Chicago Portal in layout and even in the wide variety of data provided, such as a Street Tree Census layer for Manhattan. Data layers can be visualized in Google Map and exported as KML and KMZ layers.

In comparing the number of open data "views" between Chicago and New York City (see table 2.1 below), you can see that the Chicago Open Data is perhaps more established

Table 2.1. Common Open Datasets in Chicago and New York City

These Open Datasets have been sorted by popularity (most accessed) between January 1–September 30, 2014

City of Chicago Open Data—<https://data.cityofchicago.org/>

DATASET	NUMBER OF VIEWS	NUMBER OF DOWNLOADS	% OF DOWNLOADS
Public Transit Ticket Outlets	180,205	1,088	0.6
Crimes	119,761	24,223	20.2
Potholes Patched—Last Seven Days	62,914	584	0.9
Police Stations	32,079	743	2.3
Fire Stations	28,455	1,591	5.6
Total	423,414	28,229	6.7

New York City Open Data—<https://nycopendata.socrata.com/data?type=maps>

DATASET	NUMBER OF VIEWS	NUMBER OF DOWNLOADS	% OF DOWNLOADS
Wifi Hotspots	68,252	4,052	5.9
Subway Entrances	37,258	1,592	4.3
For Hire Vehicles Locations	18,718	0	0
Basic Description of Colleges and Universities	9,004	905	10
NYC Fire Houses	8,160	905	11.1
Total	141,392	7,454	5.3

with its top 5 data layers drawing over 400,000 views while New York City has just over 140,000. Nevertheless, each of the top 5 layers for both cities relates to a direct public service that would be a strong interest to residents of those cities.

Canada

Meanwhile, north of the 49th parallel in Canada, where government information has long been bound by crown copyright, the ideas around an open data philosophy have taken a little longer to take root when compared to the United States. Prior to the arrival of Google Earth, government datasets were often trapped in top-down data management structures or data silos with only a few staff having access within different agencies. The idea of sharing such data with the public was seen as a very risky, even dangerous concept that might harm the greater public good (not to mention personal career advancement). Overall, a general distrust over what the public might actually do with the data was a common feeling that shaped Canadian public policy around data. Even if the data were made available to a user, too often it was because that person had made some valuable personal connections to data managers, or the person was asked to pay an exorbitant (and unadvertised) fee. The lack of data availability in Canada was actually so grim in the early 1990s that academics and librarians lobbied to the federal government to make some data available for research and teaching use. Up until this point, access and licensing restrictions were so severe around Canadian data, that academics had to rely on American datasets for teaching their university students. Their successful efforts led to the launch of the Data Liberation Initiative (DLI) which made a wide selection of Statistics Canada surveys and other data products available to university institutions for an annual subscription fee (Boyko and Watkins, 2011).

With the success of the DLI program and an ever-growing "born digital" environment for Canadian government data, new models began to be adopted for serving up data. For example, a fee-based model for data downloads was soon adopted for Canadian Government Geospatial data. Statistics Canada also favored a fee model for non-academic users looking for particular types of data.

Nevertheless, with the launch of Google Earth in 2005, attitudes among the federal Canadian data producers began to quickly change, and for the better. *Geobase*, a government website which provided base map data for Canada, soon began offering data in KML format and adopted an open license <http://www.geobase.ca/geobase/en/>. Even for Canadian mapping datasets which were not available to view in Google Earth, a set of KML metadata showing a map extent was available, as can be found on *Geogratis*, a Natural Resources Canada (NRCan) project which also offers up Canadian map data <http://geogratis.cgdi.gc.ca/>.

Some federal Canadian agencies also began to offer up specific datasets in KML formats to further their public outreach efforts. From an educational perspective, the availability of the Canada Lands Survey (including First Nations lands) as a Google Earth ready file <http://clss.nrcan.gc.ca/googledata-donneesgoogle-eng.php> is an especially nice bonus. This will be discussed in detail in chapter 3. Environment Canada has perhaps been the most active in releasing national data in KML and KMZ formats. For example, the agency has released a selection of datasets in KML format including the National Pollutant Release Inventory (NPRI), the Canadian Environmental Sustainability Indicators (CESI), and the Commission for Environmental Cooperation (CEC). The agency has also been active in releasing weather prediction and other meteorological datasets through its GeoMet service <http://www.ec.gc.ca/inrp-npri/default.asp?lang=en&n=1D892B9F-1>.

Inspired by the United States Open Data initiatives and the opportunities that it was generating for their own mobile app developers and digital entrepreneurs, the Canadian federal government set up its own Open Data Portal in 2013 at <http:/data.gc.ca>. Though certainly not as robust in scope or in the quantity of data offered as seen in the United States portal, it is nevertheless a good start and includes some KML files.

At the provincial level of government in Canada, open data is also slowly taking root. Ontario, Canada's most populous province, has had a "pay-as-you-go" subscription model for its geospatial data for many years, yet it was in fact the smaller province of Manitoba that lead the way with its own open data project, allowing anyone to access and download its geospatial data for no charge. Realizing that protecting and locking down data with needless firewalls was costing them too much money and inhibiting new opportunities for the province, the Manitoba government took a progressive step in setting up the "Manitoba Land Initiative" (MLI) <http://mli2.gov.mb.ca/>, which saw the province put up almost all of its provincially produced datasets onto its website online for free download in 2003. The MLI was widely heralded both politically and economically and was watched closely by data producers across the country. KML/KMZ datasets available from the MLI include administrative boundaries, cadastral, environmental layers, town plans, and much more.

Over the next few years, other provinces followed the Manitoba example, especially following the release of Google Earth. For example, British Columbia released its Open DataBC website <http://www.data.gov.bc.ca/>, which has a wide array of data layers available in KML. In 2013, Ontario finally embraced the open data movement and began releasing several datasets in KMZ/KML formats for free at <http://www.ontario.ca/government/government-ontario-open-data>. Actually, even before they had launched their open data website, Ontario had already started working with Google Earth data, serving up its geoscience data in what they branded as "OGSEarth" (Ontario Geological Survey Earth) <http://www.mndm.gov.on.ca/en/mines-and-minerals/applications/ogsearth>. This package of KML files includes layers on surficial and bedrock geology, drillholes, abandoned mines, and mineral deposits to name a few.

As in the United States, Canadian cities took their time to adopt an open data platform. In many ways the evolution of the open data movement has mirrored the growing popularity of Google Earth. The Vancouver Open Data Catalogue <http://vancouver.ca/your-government/open-data-catalogue.aspx> notes the validity of KML on its Data Formats page, and has many of its rich data layers available in this format for download. Available since 2009, the Vancouver Open Data Catalogue has been the leader of Canadian municipal open data sites, for its wealth of available data and also in its progressive open data license, which provides users with the opportunity to use and manipulate the data at no cost and without restriction. The Vancouver open data catalog includes such unique datasets as Olympic transportation plans and venues, public art locations, and even weekend playfield status updates for families.

Edmonton and Halifax are two other cities in Canada embracing open data and making their data available in the Google Earth compatible KML and/or KMZ files. Inspired perhaps by the Chicago open data portal, both Edmonton <https://data.edmonton.ca/> and Halifax <https://www.halifaxopendata.ca/> are using the same platform and model for serving up their open data. Compared with both Chicago and New York City open data websites (see table 2.1 above), both the Edmonton and Halifax sites have relatively small amounts of data currently available (see table 2.2). Nevertheless, one can see that public service related data will also be popular with these cities. Sandboxes (to assist in snow removal) are a popular public resource in northern Edmonton, while Bus Routes and Stops are already proving to be one of the most sought after data sets in Halifax. In fact, Halifax goes one step further to provide all their open data in the Esri shapefile format, which is a common format in the geospatial technology industry.

International Organizations

With the ubiquity and global popularity of Google Earth, it should come as no surprise that organizations outside of North America have also been taking advantage of Google Earth as a data-sharing and visualization mechanism. The Food and Agriculture Organization of the United Nations (FAO), for example, provides "Active Fire/Hotspot" KMLs to global citizens for download. Most impressively, these are set up as a time series providing 24 hour, 48-hour updates and a compilation of historical data <http://www.fao.org/nr/gfims/active-fire-data/kmlgoogle-earth/en/>.

Other United Nations organizations have also been using KML data to share with the public to promote greater understanding. The United Nations Institute for Training and Research (UNITAR) has collected a vast amount of geospatial data on different

Table 2.2. Common Open Datasets in Edmonton and Halifax
These Open Datasets have been sorted by popularity (most accessed) between January 1 and September 30, 2014

City of Edmonton Open Data—<https://data.edmonton.ca/>

DATASET	NUMBER OF VIEWS	NUMBER OF DOWNLOADS	% OF DOWNLOADS
Ward Boundaries	45,905	0	0
2014 Street Construction Projects	42,671	941	2.2
Sandboxes	19,793	251	1.3
2013 Municipal Wards—Elections	10,643	0	0
City of Edmonton's Properties for Sale	9,358	167	1.8
Total	128,370	1359	1.0

Halifax Regional Municipality Open Data—<https://www.halifaxopendata.ca/>

DATASET	NUMBER OF VIEWS	NUMBER OF DOWNLOADS	% OF DOWNLOADS
Bus Stops	4,243	N/A	N/A
Crime	3,038	N/A	N/A
Building Footprints	2,192	N/A	N/A
Bus Routes	1,836	N/A	N/A
Solid Waste Collection Areas	1,805	N/A	N/A
Total	13,114	N/A	N/A

member countries and can be shared with the public, including this KML set covering Sudan <http://www.unitar.org/unosat/sudan>.

Although some people may want to dismiss Google Earth as a distraction app, the free mapping applications have even helped businesses with their own needs for data visualization and analysis. A terrific example can be found at the Lloyd's Insurance Market website <http://www.lloyds.com/the-market/tools-and-resources/research/exposure-management/data-visualisation/kml-library>. As an insurance corporation working in over a dozen countries and with responsibility for billions of dollars of properties and investments, Lloyds agents use company-provided KML data to track the global impact of possible future disasters and manage investment risk in particular countries (see figure 2.4). Best of all, the data are also available to the public to download and include such layers as Earthquakes, Windstorms, Flooding, and Terrorist attacks.

Open Community/"Crowdsourced" Collections

With Google Earth's free availability and ease of use, it is not surprising to see individuals and community groups use the tool to begin KML collections of their own to share with the wider public. These "crowdsourced" collections or public gathering of datasets are much sought after for bringing added value to existing knowledge or to promote a particular cause. Perhaps the most surprising example of this in Google Earth use comes from the people behind the "OpenBible.info" whose tagline "Remix Bible data" has seen the group take up a Bible Geocoding project amongst other Web 2.0 bible initiatives. For those interested in religion and scripture, and its relationship to geography, you can visit the website <http://www.openbible.info/geo/> and access KML files from each of the books and individual chapters found in the Old and New Testaments. Each file notes the places referenced in the Bible and some include pictures. Also found at the site are historical map overlays of both ancient and modern Jerusalem.

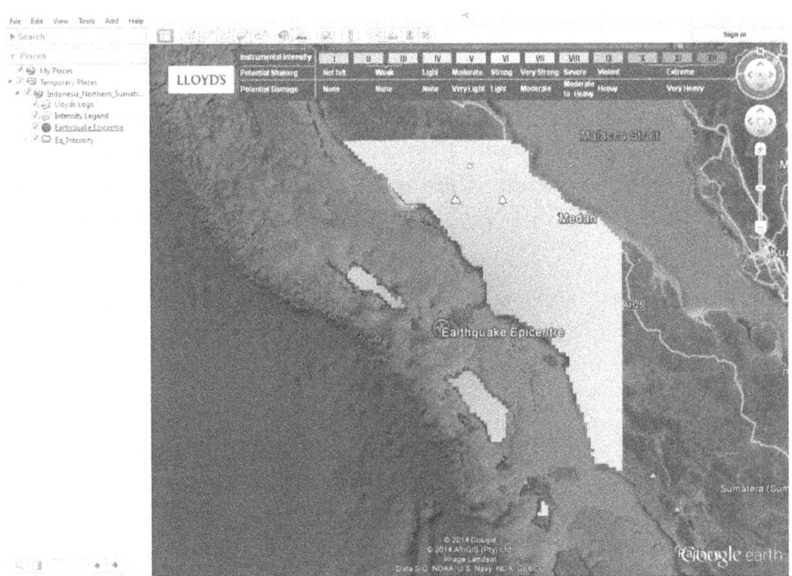

Figure 2.4. Lloyds Insurance Maps in KML: Northern Sumatra Earthquake

Source: Lloyds of London Insurance, http://www.lloyds.com/the-market/tools-and-resources/research/exposure-management/data-visualisation/kml-library> and Google Earth 7.1. Map Data: 2014 Google, 2014 AfriGIS (pty) Ltd, Landsat.

Moving into more secular areas, Google Earth has also been a powerful tool for enabling activists, journalists, and educators to raise awareness of current events, protests, disasters, and other calamities. As KML is an open format, users have been eager to use the format as a means to crowdsource information around an event and then disseminate the data to the world, in effect providing a spatial perspective of what is occurring and perhaps allowing people to see patterns or relationships between events.

Perhaps the most illustrative example of crowdsourcing and the use of KML data to share what was happening (often even in real time) occurred with Hurricane Sandy, which struck the Caribbean and the Atlantic seaboard of the United States in the fall of 2012. First responders as well as everyday citizens caught in the path of the hurricane were able to share preparation and damage information with the wider community. Often times such KML data was used to map the storm path, track the extent of the damage, organize the evacuation and relief efforts, and even practice some "citizen science" projects which would help the public learn more about Hurricane Sandy and prepare better for future storms.

Despite the storm occurring a few years ago, much of the data, including KML files are still available online. At the "Storm Center," you can still download "Sandy Briefing Products" <http://www.stormcenter.com/evcm1/sandy_briefing.html> in KML format, which includes storm path data and amounts of precipitation. Many other Hurricane Sandy resources are also available in KML when searching online. In some instances, data related to Hurricane Sandy is still being released. Perhaps the most interesting were some of the "citizen science" projects that sprung up during and in the aftermath of Hurricane Sandy. One of the most useful was the collection of water samples from the storm so scientists could study the isotopes in the water to gain a new understanding about the makeup of major storms. KML data was used to see where the samples were located. All of the water sample data including KML files can be found at Waterisotopes.org <http://wateriso.utah.edu/waterisotopes/pages/data_access/sandy.html>.

The use of KML files, Google Earth, and crowdsourcing strategies have also been seen in the "Occupy" and "Idle No More" movements taking place in recent years. In both instances activists have reached out to others through social media and mapped out the locations of their activities, and shared protest strategies. Many of these KML data are available for public download.

The Occupy Movement which began in New York City to protest against social and economic inequality soon spread around the world through the use of maps found at websites such as the Occupy Wall Street website, which both maps members and provides an embedded text service to rapidly communicate <http://occupywallst.org/attendees/>.

"Idle No More" was a movement in Canada to protest Canadian government policies toward native land claims and lack of attention to native issues. Beginning in late 2012, the protests led to a series of teach-ins, flash mobs, road blockages, and rallies, which soon spread from Canada to the United States and other countries. Using social media and tools such as Google Maps, protesters were able to communicate and share event information and locations as KML files. Just searching online for "Idle No More" will bring up user generated KML files such as this one: <http://idlenomore.tumblr.com/events>.

In fact, one of the best examples of activists using crowdsourced KML data to promote their causes can be found at the Act MTL website <http://actmtl.com/>, which chronicles a history of protest in Montreal, including details and locations that users can open in Google Earth.

Other sources that might often be overlooked for "protest" data are "think tanks" or research institutes attached to a particular social issue or political cause. The Cline Center for Democracy located at the University of Illinois, for example, has a surprising amount of KML data related to civil unrest demonstrations, human rights violations, and coup d'états. The Cline Center, through its SID and SPEED data mining projects, went through global newspaper archives from 1945 to the present day to collect data on political events that can be mapped to provide an historical snapshot of political change and public protest. This data can be located at <http://www.clinecenter.illinois.edu/research/speed-data.html>.

While many community and activist websites will post KML files of their activities, perhaps the best place to start looking for KML data around particular newsworthy events is at the "Geocommons" portal <http://geocommons.com/>. The Geocommons portal is one of the largest repositories of user generated geospatial data. The portal not only allows users to access and download geospatial data, including KML files, but it also allows users to make maps online using data that have been uploaded to its cloud servers. It offers a free sign up and also has special feature project datasets including Hurricane Sandy, the Occupy Movement, and many others. One particularly rich source of data that can be found here are geotagged "tweets" from Twitter. These can be especially useful for mapping public reaction to real-time events or news reports. Geocommons, for example, has geotagged Twitter data from the 2012 United States presidential election debates and can be mapped to see how statements and debate points scored with the American public in different parts of that country and even around the world.

Key Points

Because of its universality and ease of use, more and more libraries are utilizing Google Earth as a gateway to digitization projects and geospatial data. With the growing popularity and increasing use of KML file formats, the general public is becoming more aware of the potential of Google Earth as a data visualizer. Libraries are experimenting with using Google Earth as a replacement for the traditional library catalog as well as a platform for creating easy-to-use pathfinders and for discovering and downloading library collections. Many are offering data files in Google Earth supported file formats, broadening the user base from traditional GIS users to those who weren't aware they were using spatial resources.

With the increased interest in KML files, libraries and government organizations have been putting their efforts into opening their data, sharing them with the public, and teaching users how to use the files in Google Earth. Although the library may not be an exclusive data repository, the teaching and learning resources it offers are second to none.

- Google Earth is becoming a popular interface for viewing pathfinders for hard to find print material
- Many libraries are using Google Earth as a discovery tool for library resources
- More and more special online projects are being hosted using Google Earth
- Libraries are discovering and managing KML files and open data collections from government organizations

The next chapter will focus on the many different disciplines that can use Google Earth in the classroom. Examples of ways that Google Earth has been used by teachers and students will be reviewed, followed by suggested classroom activities that would be appropriate for elementary grades (public library system librarians) and postsecondary education (academic librarians).

References

Bacharach, Sam. 2008. "OGC Approves KML as Open Standard." *Open Geospatial Consortium Press Release* April 14. http://www.opengeospatial.org/node/857.

Boyko, Ernie, and Wendy Watkins. 2011. "The Canadian Data Liberation Initiative: An Idea Worth Considering?" *International Household Survey Network Working Paper No 006*. http://www.ihsn.org/HOME/sites/default/files/resources/IHSN-WP006.pdf.

Morris, Steven P. 2006. "Geospatial Web Services and Geoarchiving: New Opportunities and Challenges in Geographic Information Services." *Library Trends* 55, no. 2. 285–303.

Tauberer, Joshua. 2014. "Open Government Data: The Book." Second Edition. http://opengovdata.io/.

Wikipedia. 2014. "Open Data." Wikipedia, the Free Encyclopedia. Accessed August 4. http://en.wikipedia.org/wiki/Open_data.

CHAPTER 3

Google Earth: A Technological Tool for Teaching and Learning

IN THIS CHAPTER

- ▷ Introducing Google Earth into the classroom
- ▷ Building lesson plans with Google Earth
- ▷ Google Earth used in K-12 classrooms
- ▷ Teaching resources for Google Earth
- ▷ Google Earth for higher education
- ▷ Google Earth and Earth Science: opportunities for teaching and learning assessment

GOOGLE EARTH HAS BECOME A POPULAR INSTRUCTIONAL TOOL inside and outside of educational settings, allowing users to explore different subject matters within standards-based, inquiry-driven lessons. When accompanied with appropriately designed learning and support materials, Google Earth can be used to foster spatial thinking skills and help develop critical analytical skills. K-12 teachers, college and university instructors, as well as many librarians are now using Google Earth to teach geospatial, environment, geography, science, and other subject specific lessons. This chapter will examine how educators and students use Google Earth inside and outside of the classroom. Specific activities and lesson plans will also be discussed to guide readers in preparing their own Google Earth teaching strategies.

Introducing Google Earth into the Classroom

The advancement of geospatial technology has been leading to increased ease of use and access to online mapping applications. This has naturally been impacting and changing the way some subjects are taught, and the way teachers and students are learning. Engaging with the tools, collaborating with others, and sharing knowledge in an online environment is shifting the paradigm of skills acquisition away from the physical classroom and study environment. Students are eager to explore and engage with online resources, to make a difference by dabbling with crowd sourcing, and "learning by doing." Incorporating technology in learning activities has been shown to improve student motivation and increase achievement (Andrews, 2006). Instructors are encouraged to accommodate the changing times and teach students using media and technology that the students have grown up with. This may be a challenge at times since some teachers are not comfortable using new technology, much less mapping software. Few teachers are willing to delve into content areas or use technology that they are uncomfortable with, making spatial literacy and GIS training crucial for both teacher and student.

Since the development of Web 2.0 several multi-literacy models have emerged, redefining literacy to include virtual communities, collaborative processes, and information sharing (Whittingham et al., 2013). Integrating Web 2.0 technology into the classroom and throughout instruction will provide students with the foundations necessary to learn the content and the opportunity to go beyond traditional learning objectives. The virtual globe offers an understanding of more than just the subjects of geography, science, history, economics, sociology, etc. It offers a spatial representation of our planet, visually representing information found on the Earth, on a map, in a textbook and online, and when combined with collaboration tools, promotes and instills spatial thinking, a critical skill for people living in the twenty-first century (National Research Council, 2006).

Google Earth can aid teaching and learning in various capacities. Students can learn about fundamental geography, like river deltas, topography, floodplains, deforestation, natural disasters, urban sprawl, and historical events. Math students have been known to learn about polygons by visually analyzing "The Pentagon," and learning about solids by observing "The Pyramids of Egypt" (Butler, 2008). Students can use the tools for inquiry-driven, problem-based, and exploratory learning using features like the placemarks, measuring distances between two points, or collecting information from the layers. Hyperlinks can be added into the placemarks and can be used to lead students to lesson files housed on other web servers. More recently, users have become interested in mapping stories, such as locations of routes, events, and narratives. Many traditional resources that are read and analyzed can be geospatially referenced and plotted onto a map to determine their geographic relevance. For example, teachers can add relevance and make literature come alive by visualizing the settings of books through the use of maps, photos, YouTube videos and photographs (Jenny, 2011). It is estimated that 80 percent of all data available today can be geocoded or georeferenced (National Research Council, 2006). This of course increases the need for spatial literacy skills, something that can begin developing in elementary school children.

Finding websites that offer KML files for download further enhances the learning experiences, offering students images, statistics, and graphics on a large number of topics. For example, there is a wealth of environmental information available in KML file format including agricultural resources, forest fires and their impact, ozone measurements, as well data on the cryosphere and hydrosphere. The public can also find KML files related to population density, earthquake activity, snow cover, volcano activity, plate boundaries, elevation and bathymetry, temperature, relative humidity, rainfall, snowfall, vegetation index, water quality, soil moisture and temperature, biomass, and many others (Rakshit and Ogneva-Himmelberger, 2008). KML files not only introduce students to GIS technology and the concept of geospatial files, they also plant the seed for further discovery of GIS data and software programs. With the ability to convert a GIS file into a KML file format, and as the delivery of data in KML format increases, the information that can be viewed, mapped, and analyzed in Google Earth is essentially limitless. Educators encourage their students to use multiple software tools that include geocoding, creating html files, and using open source GIS or online tools to convert from CSV to KML files. In one post-secondary class, for example, the students were asked to collect earthquake data from the USGS and import them into Google Earth. The year, month, day, and time of the earthquakes were also imported, and used to create time animation demonstrating the sequence of earthquakes and aftershocks (Guth, 2012). These types of hands-on learning experiences invite students to make observations, reflect on their findings, ask questions from a variety of perspectives, and formulate their own hypothesis (Almquist, Blank, and Estrada, 2012).

Integrating Google Earth in the K-12 Classroom

According to the National Middle School Association (NSMA), also known as the Association for Middle Level Education (AMLE), one of the fundamental beliefs is that teachers and students should actively engage in learning (Whittingham et al., 2013). Integrating Google Earth within the K-12 curriculum allows educators to create interactive classroom activities that are creative and that deepen understanding of the world around them. Many instructors look for meaningful ways to connect Google Earth with texts, videos, photographs, and other resources in a variety of subjects. Most common and easily adaptable is for geography and science students, but Google Earth lesson plans have also been created for literature, social studies, math, and art students. As a digitally-based resource, Google Earth also supports the International Society for Technology in Education (ISTE) as it provides productivity tools to collaborate and communicate, and locate, evaluate, and collect information from a variety of sources.

Geography Education

Technological developments in GIS and Google Earth have contributed to the need for spatial literacy in geography and related social studies education. The first three Essential Elements of the eighteen National Geography Standards framework define the world in

> Effective teachers model and apply the ISTE Standards for Students as they design, implement, and assess learning experiences to engage students and improve learning. The five standards are available from <http://www.iste.org/docs/pdfs/20-14_ISTE_Standards-T_PDF.pdf>.

VOLUNTARY NATIONAL STANDARDS FOR GEOGRAPHY

Voluntary national standards for the study of geography were agreed upon in 1994 in the United States by a consensus of educators, parents, and other interested citizens. These standards outline what students should know and be able to do in geography, and are organized into six "essential elements": (1) the world in spatial terms (location); (2) places and regions; (3) physical systems; (4) human systems; (5) environment and society; and (6) the uses of geography (Geography Education Standards Project, 1994). Canada has also adopted these standards.

spatial terms <http://alliances.nationalgeographic.com/detail/the-national-geography-standards/edn7EBAD664C732E0D98>, enforcing the need to understand how to use maps and other geographic tools and how to analyze spatial representations of people, places, and environments on the Earth's surface (Geography Education Standards Project, 1994).

Incorporating the National Geography Standards into the K-12 geography classroom is not a difficult task when using geography-based technology. Effective use of technology, like Google Earth in the classroom and in course projects, has been examined by a number of researchers. Many studies have been conducted examining the types of lessons that could be developed, and their benefits, while testing student retention, their gained skills, enhancement of their spatial thinking skills, and their overall enjoyment of the lessons.

In one study, a teacher-professor team worked together using Google Earth to teach elementary geography, specifically focusing on landforms, while teaching to read maps and globes. Students learned how to use Google Earth to discover their local area, and then to virtually visit landforms around the world. Students used Google Earth's embedded layers like photographs, air photos, and satellite images to explore and make observations, and were then tested on their new knowledge. Their lesson plan met the state and national standards in the K-6 curriculum while providing the students with an engaging experience that deepened their understanding of their place in the world (Britt and LaFontaine, 2009).

In a study by Doering and Veletsianos (2008), investigators sought to understand students' experiences learning geography when using Google Earth within their classroom. Students were given the task to study issues surrounding global climate change and their sense of place within their home state and the state of Alaska using real-time data. Students used Google Earth to explore their home and surrounding neighborhoods and communities, making note of the designs of streets, familiar places like local shops, the shapes and sizes of the houses and rooftops, and walking paths. From this, students were able to appreciate the spatial nature of their familiar surroundings in relation to the larger world around them. The students were also asked to explore the unfamiliar state of Alaska, adding GIS data into Google Earth, which allowed them to further explore the landforms and population of the state. In all their work, students were encouraged to create and share the KML files with others.

The results revealed that Google Earth assisted learners in developing a sense of place using pre-existing and newly acquired data, motivated students to take the initiative to explore geographic locations that were beyond the scope of the project, and assisted them in understanding the geography of the region. Students were excited and motivated to use Google Earth inside and outside of the classroom (Doering and Veletsianos, 2008).

Bodzin, Anastasio, and Kulo (2014) developed two instructional middle school Google Earth modules to promote understandings of earth and environmental science concepts. One module was designed to teach students about land use change concepts and aerial photography and interpretation. Students studied the photographs to identify man-made and physical geography features, as well as land use patterns. The second module examined energy sources and had students locate solar power plants from around the world while virtually investigating potential future sites. Using a combination of aerial photography, measurement tools, and KML files, the students also looked at wind farms and learned about tidal power and hydroelectric dams. These exercises followed specific design principles which included motivating contexts and personally relevant and meaningful examples.

Science Education

Science focuses on the analysis and synthesis of information, and when combined with Google Earth, it can provide teachers and students with an opportunity to use active learning strategies. Geospatial explorations are common in the scientific field, so Google Earth is a familiar resource in many science classrooms. The Science Education Resource Center (SERC) at Carleton College <http://www.serc.carleton.edu> lists seven broad categories in which Google Earth can be used in the classroom:

1. To support hands-on inquiry by students in computer classrooms
2. As a basis for homework assignments
3. For dynamic presentations during class lectures
4. For inquiry during class presentations
5. To create imagery and maps for PowerPoint, Word, and other presentation tools
6. As a data discovery, organization, and distribution tool for research projects
7. To enrich discussion of an issue that arises spontaneously during an informal classroom discussion

Several researchers and educators have incorporated these opportunities into the classroom. Popular science topics to teach using Google Earth include plate tectonics, earthquakes, sea level changes, glaciers, climate changes, ecology, and hydrology.

One researcher examined how Google Earth could be used in elementary school children to promote investigative learning about their watershed and their pond's ecosystem. A fourth grade after-school science club in Pennsylvania was formed to improve the students' understanding of their local watershed by using the resources available to them in Google Earth to investigate a pond system in their local schoolyard. The science club met weekly after school throughout the school year with the primary focus to investigate the ecology of the pond over three seasons and then share their findings with others.

Throughout the year the students explored the biotic and abiotic components of the pond, and learned about basic water quality parameters. They used web maps and Google Earth to learn about their watershed and to display images of the school, pond, and the creek that feeds into the pond, as well as trace the path of the river. By visualizing how the pond flows, students came to understand how ponds are part of a greater watershed area, connecting to creeks, rivers and ultimately the ocean (Bodzin, 2008).

THE SCIENCE EDUCATION RESOURCE CENTER

The Science Education Resource Center (SERC) works to improve education through projects that support educators. Although their work has a particular emphasis on Science, Technology, Engineering, and Mathematics (STEM) education, they work with educators across a broad range of disciplines and at all educational levels. Their work is funded primarily through National Science Foundation grants.

Literacy Education

Google Earth is becoming a common tool to use to connect literature to a geographic place. Accompanying traditional books, students can use the mapping application to find towns, placemarks, and other points of interests discussed in the literature. Additionally, students are able to visualize physical geographic landforms described in texts. For students who are unfamiliar with deserts, volcanoes, jungles, and other location specific land areas, they can use Google Earth to virtually visit and examine the physical landforms, providing them with a better idea of what they look like.

Characters can also become more personalized as readers travel to the locations with them, understanding where they had been and the journey they had taken. The popular children's book, *Flat Stanley*, for example has been updated from its original 1964 publication to offer students clues in Google Earth to accompany the scavenger hunt-type book. Students follow Flat Stanley's journey around the world, learning about new locations around the world while doing so (Misegadis, 2011).

Another interesting project is by educator Jerome Burg called *Google Lit Trips* <http://www.googlelittrips.com>; it offers a collection of journeys that provide meaningful context for children's literature. Designed to supplement specific literature in order to gain a deeper understanding of the books they are reading, each trip is an interactive multimedia experience created with Google Earth and shared via KMZ files. Students can virtually fly from one story locale to another and zoom into the setting details, travelling and discovering from different angles and heights (Jenny, 2011).

Google Earth serves as a wonderful accompanying resource for students reading textbooks, traditional literature, or modern novels. Visualizing the settings of books assists in mentally building up the characters, as well as aids in interpreting the physical descriptions of landforms and buildings. Reading experiences used in conjunction with Google Earth encourage critical thinking skills in literacy such as interpreting, analyzing, comparing, contrasting, and explaining (Castek and Mangelson, 2008).

Lesson Plans

Creating lesson plans for K-12 courses is a relatively simple task because of the abundance of resources available in Google Earth and the wide variety of topics that are covered in the courses. Lesson plans and ideas for assignments are available online via educational websites as well as Google Earth's website <http://www.gelessons.com/lessons/>, which offers a large collection of lesson plans on a variety of topics. The following offers a number of resources that provide complete ready-to-use lesson plans that follow national teaching standards.

Google Earth for Educators

Google Earth for Educators <http://sitescontent.google.com/google-earth-for-educators/classroom-resources/lesson-plan-library> offers a Google Earth Lesson Plan Library that includes classroom-ready lesson plans for a variety of subjects like geography, history, and science covering topics like the Renaissance, the solar system, animal habitats, food chains, and plate tectonics, to name a few. Each lesson incorporates State and National Educational Technology Standards, and includes the instructional goals and objectives, the description of the activity and procedure, and links to resources.

A number of classroom resources are available from <http://sitescontent.google.com/google-earth-for-educators/classroom-resources/projects-for-my-subject>, which offer Google Earth activities for a number of subject areas such as environmental and earth sciences, history, social studies, humanities, art history, architecture, English and literature, and math. Each project is marked as easy, average, or advanced, depicting the assignment's difficulty level. An example of a history assignment has students exploring the power of photographs and their impact on history. Photographs are overlaid in Google Earth and students are asked to consider the significance of the photo and the significance of the location where the photo was taken. Several more collaborative-like questions are asked as students research the photographs and find others that have influenced historical events. The website invites teachers to submit their lesson plans, and it also invites students to submit their completed projects.

The National Association of Geoscience Teachers (NAGT)

The National Association of Geoscience Teachers (NAGT) has a well-developed website <http://serc.carleton.edu/NAGTWorkshops/google_earth/index.html> that offers a large number Google Earth-related resources including teaching activities about streams and glaciers, earthquakes, tsunamis, erosions, weathering, and so much more. The NAGT offers a fabulous resource for learning how to teach using Google Earth, including information about:

- Google Earth basics
- Books and online resources for using Google Earth
- Potential uses of Google Earth for both students and teachers
- How to get started teaching with Google Earth
- Working with KML
- Ready to use classroom activities

There are approximately forty-five ready-to-use classroom activities contributed by teachers and professors covering a wide variety of geological themes. An example of an assignment has students using Google Earth to investigate a variety of fluvial and glacial features. First, they are asked to find an example of both a meandering and braided river and cut-and-paste the image into their assignment. They also need to trace the path of the river to see where it ends. Next, they are given three latitude/longitude coordinates and are asked to determine the river type, channel width, floodplain width, gradient, etc. In the last part, they are given the latitude/longitude coordinates of two valleys. They use the terrain and tilt features of Google Earth to determine whether the valley is V- or U-shaped. They then decide what sort of processes are responsible for the valley's shape. This assignment is available from <http://serc.carleton.edu/NAGTWorkshops/intro/activities/23748.html>. A screen capture of a typical assignment can be seen in figure 3.1.

Real World Math

Real World Math <http://www.realworldmath.org/> offers students and educators a large collection of math activities for Google Earth, including lessons that offer exploratory and project-based learning. There are over thirty activities, videos, and instructional tutorials for grades five and up. The lessons are organized by content, such as math concepts, measurement, project-based, exploratory, and space lessons. Examples of some of the lessons available include estimation, exchange rates, line graphs, scientific notation, time zone problems, and so forth. Each lesson has an associated KMZ file, which the student will need to download and then interact with using Google Earth. The KMZ files consist of placemarks with instructions, information, and images about the content covered. Real World Math also offers a series of video tutorials that demonstrate a particular skill used in Google Earth, such as adding placemarks, opening KMZ files, and using the ruler tool.

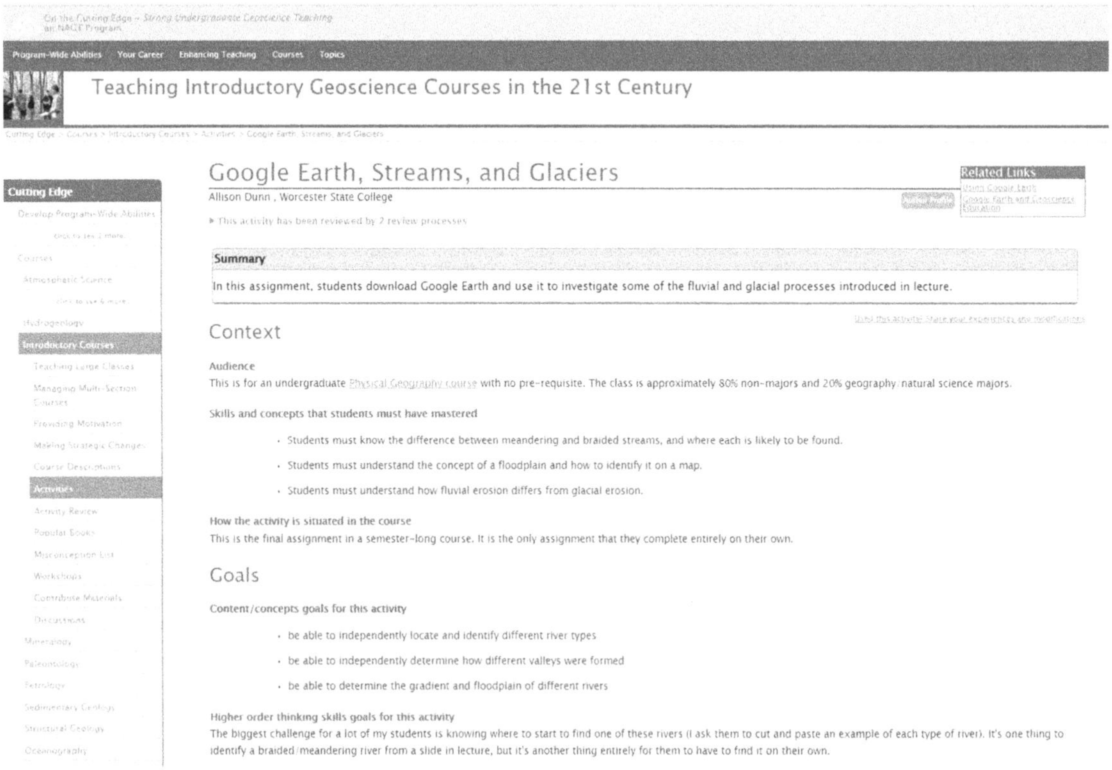

Figure 3.1. The National Association of Geoscience Teachers (NAGT) sample Google Earth classroom assignment.

Source: <http://serc.carleton.edu/NAGTWorkshops/intro/activities/23748.html>

Geographic Alliance of Iowa

The Geographic Alliance of Iowa (GAI) is composed of K-12 teachers, college and university professors, and individuals from a variety of other educational sectors who are concerned with improving geographic instruction in Iowa. GAI's mission is to help ensure K-12 graduates are geographically literate through the provision of standards-based online resources. Their website <http://www.uni.edu/gai/map-resources> provides a large number of Google Earth lesson plans, curricular material, and other teaching resources. Lesson plans are available for grades K-12 and can be searched using a pull-down feature. Users can search for lesson plans by discipline/subject (i.e., geography, sociology, science, reading, music, etc.), by theme (i.e., climate, population, urban, etc.), and by several standards (i.e., Iowa Core 21st Century Standards, Iowa Core Social Studies Standards, etc.).

There are a large number of Google Earth lessons available that cover a wide variety of topics. Table 3.1 lists some of the topics and associated links to the lesson plans. The sidebar is a sample assignment taken from <http://www.uni.edu/gai/lesson-plans-and-curricular-materials/9-12/introduction-egypt>, showing how the features in Google Earth are being used to learn about cities.

There are many other online resources educators can go to for Google Earth-based assignment ideas and lecture material. Table 3.2 offers a sampling of some of the various resources discovered online, organized by subject area.

Google Earth for Higher Education

As discussed through the first part of this chapter, Google Earth has played a prominent role in promoting Science Education and Spatial Literacy concepts in many K-12 classrooms. In many instances, teachers have been eager to try out the new technology as part of a broader initiative to better engage young students with the world around them. In the post-secondary environment of learning, Google Earth has also been adopted, although perhaps a bit more slowly at first, especially among educators who were accustomed to high-end and powerful GIS applications such as GRASS and ArcGIS. The initially slow take up of Google Earth in the post-secondary environment was noted by Professor Simon Haslett at the Centre for Excellence in Learning and Teaching at the University of Wales. Checking the powerful Web of Knowledge database at the end

Table 3.1. Geographic Alliance of Iowa Google Earth Lesson Plans

SUBJECT	LINK TO LESSON PLAN
Geography	http://www.uni.edu/gai/lesson-plans-and-curricular-materials/9-12/effects-monsoon-seen-through-google-earth http://www.uni.edu/gai/lesson-plans-and-curricular-materials/9-12/introduction-egypt http://www.uni.edu/gai/lesson-plans-and-curricular-materials/9-12/virtual-vacation http://www.uni.edu/gai/lesson-plans-and-curricular-materials/9-12/geography-rwandas-genocide
History	http://www.uni.edu/gai/lesson-plans-and-curricular-materials/9-12/using-google-earth-track-carnegie-steel-us-history http://www.uni.edu/gai/lesson-plans-and-curricular-materials/6-8/ancient-civilizations-review
Cartography	http://www.uni.edu/gai/lesson-plans-and-curricular-materials/6-8/mapping-crusades
Planning	http://www.uni.edu/gai/lesson-plans-and-curricular-materials/3-5/urbansuburbanrural-communities

INSTRUCTIONAL SEQUENCE/PROCEDURE

1. Assign appropriate reading of text and an introductory take-home quiz about Egypt in anticipation of day 1 of the lesson. The quiz will not be graded, but will be used as a tool to gauge how well students understand the reading and their previous knowledge about Egypt.
2. (Start of lesson) Have students determine the absolute location of Egypt and various Egyptian cities using Google Earth.
3. Discuss with students the relative location of Egypt using the outline maps. Have students refer to the surrounding countries and bodies of water.
4. In reference to place, introduce to students the concept of "physiological density." Have students observe that the population distribution within Egypt lies primarily along the banks of the Nile River.
5. Have students compute (both) the population and physiological density of Egypt. The students will discuss what each measurement implies about Egypt (how different conclusions about density will be drawn based on the measurement tool). Online sources of information (i.e., population, total land area, arable land area) must be credible (i.e. CIA World Factbook).
6. Have students research and discuss why Egypt is included in the North Africa/Middle East region. The characteristics to be discussed should include the following: (1) located near or in deserts and mountains, (2) a population that practices one of the major monotheistic religions, (3) possessing oil reserves, (4) a need for freshwater resources, and (5) a long history of external influence. Refer to hard copy maps when needed.
7. Present students with a short lecture about the role that, in the past and present, the Nile River plays in facilitating movement within the developed region of Egypt.
8. Using Google Earth and other websites, students will treat the Aswan Dam as a case study of how the Egyptians have chosen to interact with the environment (specifically, the Nile River). Just such interaction (the need to build dams and canals) is what caused the first government to spring up along the Nile River. Students will discuss positive outcomes (i.e. irrigation, water supply, flood control) related to the control of the Nile. They will also be asked to explore unintended consequences (i.e. strategic vulnerability of the Aswan Dam, water-borne diseases) related to the attempt to control the Nile. They will record both the pros and cons in their notebook/portfolio.
9. The introduction will close with a discussion of the problems facing the present-day Egyptian government (current events discussion).

of March 2009 (almost four years after Google Earth's public release), Professor Haslett found a mere 38 research articles with Google Earth in the article title, and only 123 more articles which identified Google Earth as a research topic across the domains of the Sciences, Social Sciences, and Humanities. Moreover, in surveying 56 undergraduate students at his university around the same time period, Professor Haslett was also surprised to find that only 23 percent had used Google Earth for academic work, and only

Table 3.2. Google Earth Teaching Resources for K-12 Education

SUBJECT	PROJECT TITLE
Classics	**Google Earth Scavenger Hunt for Elementary School**
	URL:http://middleschoolcomputerlessons.wordpress.com/2012/01/31/middle-school-computer-lessons-google-earth-scavenger-hunt/
	A Google Earth Introduction to Remote Sensing
	URL:https://www.google.ca/#q=a+google+earth+introduction+to+remote+sensing
	Gone Google
	URL:https://plus.google.com/photos/110622333232854759880/albums/5076777851562286417?banner=pwa
Science	**Introductory Biology**
	URL:http://google-latlong.blogspot.ca/2008/02/putting-google-earth-to-test.html
	Energy Consumption Rates Across the USA and the World
	URL:http://serc.carleton.edu/sp/library/google_earth/examples/energy.html
	Solar System
	URL:http://sitescontent.google.com/google-earth-for-educators/classroom-resources/lesson-plan-library
	Food Chains
	URL:http://sitescontent.google.com/google-earth-for-educators/classroom-resources/lesson-plan-library
	Modeling Sea Level Change on the Gold Coast
	URL:http://www.juicygeography.co.uk/sealevel
Geography	**Streams and Glaciers**
	URL:http://serc.carleton.edu/NAGTWorkshops/intro/activities/23748.html
	Upper Yellowstone River Drainage Basin
	URL:http://serc.carleton.edu/NAGUpper Yellowstone River Drainage BasinTWorkshops/geomorph/activities/24770.html
	The Color Run Project
	URL:http://www.uni.edu/gai/lesson-plans-and-curricular-materials/9-12/color-run-project
	Exploring Time Zones—A Travel Experience
	URL:http://gelessons.com/lessons/newlessonfiles/timezones.html
	More detailed lesson plans offered by the same site in similar manners
	URL:http://www.gelessons.com/lessons/
	Brahmaputra Flooding Lesson Plan
	URL:http://www2.geog.soton.ac.uk/users/TrevesR/TOE/lesson_plans.html
Anthropology	**The Diamond Trade**
	URL:http://www.juicygeography.co.uk/diamonds.htm
	Outbreak
	URL:https://docs.google.com/document/d/1Q91fHUw5yg5zpbFL02y-bV11R5VtzD42I05NHwtWjol/preview#bookmark=id.c2b1ef2da833

(continued)

Table 3.2. *(Continued)*

SUBJECT	PROJECT TITLE
Archaeology	**Walking with the Emperors: Exploring Imperial Rome Using Google Earth's Ancient Rome 3D**
	URL:https://www.google.ca/#q=Walking+with+the+Emperors:+Exploring+Imperial+Rome+Using+Google+Earth+Ancient+Rome+3D&spell=1
	Ancient Egypt with Google Earth
	URL:http://www.myfreshplans.com/2010/10/ancient-egypt-with-google-earth-2/
Forensics/CSI	**Float Along with Huck and Jim**
	URL:http://www.uni.edu/gai/lesson-plans-and-curricular-materials/9-12/float-along-huck-and-jim
	CSI: Cemetery Scene Investigators
	URL:http://www.evalamar.com/GPS/GPS-activities.htm
Earth Science	**Impact of Climate Change**
	URL:http://sitescontent.google.com/google-earth-for-educators/classroom-resources/lesson-plan-library/impact-of-climate-change
	Effects of the Monsoon as Seen Through Google Earth
	URL:http://www.uni.edu/gai/lesson-plans-and-curricular-materials/9-12/effects-monsoon-seen-through-google-earth
	The Boxing Day Tsunami
	URL:http://serc.carleton.edu/sp/library/google_earth/examples/boxing_day_tsunami.html
	San Francisco: visualizing a safer city
	URL:http://www.juicygeography.co.uk/googleearthsanfran.htm
Environmental Management	**Hawaii Volcano Hazard**
	URL:http://googleearthdesign.blogspot.ca/2008/11/lesson-plan-hawaii-volcano-hazard.html
	Environmental impact of oil production in Alaska
	URL:http://serc.carleton.edu/sp/library/teachingwdata/examples/25102.html
	Selecting Sites for Renewable Energy Projects
	URL:http://serc.carleton.edu/sp/library/google_earth/examples/renewable_energy.html
	An Introduction to Land Use and Land Cover
	URL:https://www.google.ca/#q=An+Introduction+to+Land+Use+and+Land+Cover+google+earth
Math	**Real World Math—Using Google Earth in Math Curriculum**
	URL:http://www.classroom20.com/forum/topics/649749:Topic:127632
	RealWorldMath: A collection of similar projects with lesson details
	URL:http://www.realworldmath.org/
	Google Earth Lesson Plans for Math
	URL:http://www.myfreshplans.com/2010/09/google-earth-lesson-plans-for-math/
	Math Teaching: A collection of similar projects with lesson details
	URL:http://teachingwithgoogleearth.wikispaces.com/

SUBJECT	PROJECT TITLE
History	**Ancient Civilization Review** URL:http://www.uni.edu/gai/lesson-plans-and-curricular-materials/6-8/ancient-civilizations-review
	Using Google Earth to Track Carnegie Steel from U.S. History URL:http://www.uni.edu/gai/lesson-plans-and-curricular-materials/9-12/using-google-earth-track-carnegie-steel-us-history
	Introduction to Egypt URL:http://www.uni.edu/gai/lesson-plans-and-curricular-materials/9-12/introduction-egypt
	Mapping the Crusades URL:http://www.uni.edu/gai/lesson-plans-and-curricular-materials/6-8/mapping-crusades
	Postcards from the Past with Google Tools URL:http://sitescontent.google.com/google-earth-for-educators/classroom-resources/projects-for-my-subject
	Using Google Earth to Teach Pearl Harbor URL:http://www.atsolutions.info/google_earth/index.asp
	Drake's Circumnavigation URL:https://sites.google.com/site/gtaukge/case-studies/drakes-circumnavigation
Social Studies	**Urban / Suburban / Rural Communities** URL:http://www.uni.edu/gai/lesson-plans-and-curricular-materials/3-5/urbansuburbanrural-communities
	Virtual Vacation URL:http://www.uni.edu/gai/lesson-plans-and-curricular-materials/9-12/virtual-vacation
	The Geography of Rwanda's Genocide URL:http://www.uni.edu/gai/lesson-plans-and-curricular-materials/9-12/geography-rwandas-genocide
	Google Earth in the Social Studies Classroom URL:http://cnx.org/contents/186342a6-4e1a-491e-ae7b f3a257c85a23@1/Google_Earth_in_the_Social_Stu
	Exploring the Great Monuments of the World URL:http://64.233.179.110/educators/learning_materials/Earth_Lesson_HighSchool_Geography.pdf
Economics	**Google Earth and the Financial Crisis** URL:http://geographyfieldwork.com/GoogleEarthTeaching.htm

11 percent were taught the program for use in coursework. Noting to other post-secondary instructors that they should not assume that students have prior knowledge of Google Earth, Haslett encouraged its adoption in more higher education activities across disciplines (Haslett, 2009).

In the five years since Professor Haslett's research, Google Earth has clearly found a greater acceptance in university research and teaching. Performing the same searches in the Web of Science (formerly Web of Knowledge) on November 4, 2014, produced 239 scholarly papers with Google Earth in the title, and 981 article hits when doing a Topic search. What is especially interesting is seeing the broad range of research and teaching topics across disciplines that Google Earth has encouraged with its easy availability and functionality for learners.

This part of chapter 3 will highlight uses of Google Earth pedagogy in the post-secondary classroom and present teaching models developed by higher education instructors. Unlike K-12 lessons, Google Earth labs in university and college courses have put greater emphasis on learning outcomes, with many looking at ways that incorporate learning assessment models such as "scaffolding" and "Bloom's Taxonomy," which will be of particular interest to academic librarians who teach.

Another aspect of teaching with Google Earth in the post-secondary classroom is the broad diversity of subjects in which you can introduce the application. With academic librarians working in multidisciplinary subjects and active teaching environments each day, Google Earth can be a terrific tool to engage students, no matter how many years of post-secondary learning they have accomplished or what their major of study is.

As demonstrated in the case studies below, many instructors who had never thought about maps in their classrooms before have been quite proactive on using Google Earth to engage their students in new ways. Along with Geography and Environmental Science-based curricula, Google Earth is being used in the teaching of Aboriginal and First Nations Studies, Archaeological methods, Historical Studies, Language Learning, and even Psychology to name just a few examples.

Teaching Aboriginal and First Nations Studies using Google Earth

As discussed in chapter 2, the availability of free "open data" in the United States and Canada has helped spur interest in local mapping activities among citizens and the development of spatial literacy initiatives in schools and among other groups. The "Canada Lands Overlay" in particular is one "open" dataset that is available as a Google Earth layer (see figure 3.2) and has been utilized by many students in Canadian Geography classes <http://clss.nrcan.gc.ca/googledata-donneesgoogle-eng.php>. The overlay file consists of administrative boundaries and cadastral parcels of First Nations Reserves, National Parks, and territories. With hundreds of reserves in Canada, the overlay file is large but the data is arranged by province, and each reserve and accompanying parcels are appropriately labelled, no matter the size.

One typical university assignment that utilizes this data involves students researching Native Land Claims on the Government of Canada website <https://www.aadnc-aandc.gc.ca> and then looking up that particular claim and finding its location in Google Earth using the "Canada Lands Overlay" KML file. Students were asked to do further research and find media articles and video stories about the land claims, photos of the disputed land, plus advocacy group websites in support of the claims and/or those opposed. All of these elements were then put into a Google Earth mashup as part of their assignment and submitted to their instructor.

Teaching Archaeology using Google Earth

As "location" plays an intricate part in the discipline of Archaeology, it is no surprise to learn that Google Earth is used in many Archaeology labs and classrooms for both informal learning and more formal assignments. One such case is documented by Professor Elaine Sullivan (2011), who teaches in the Digital Cultural Mapping Program at UCLA.

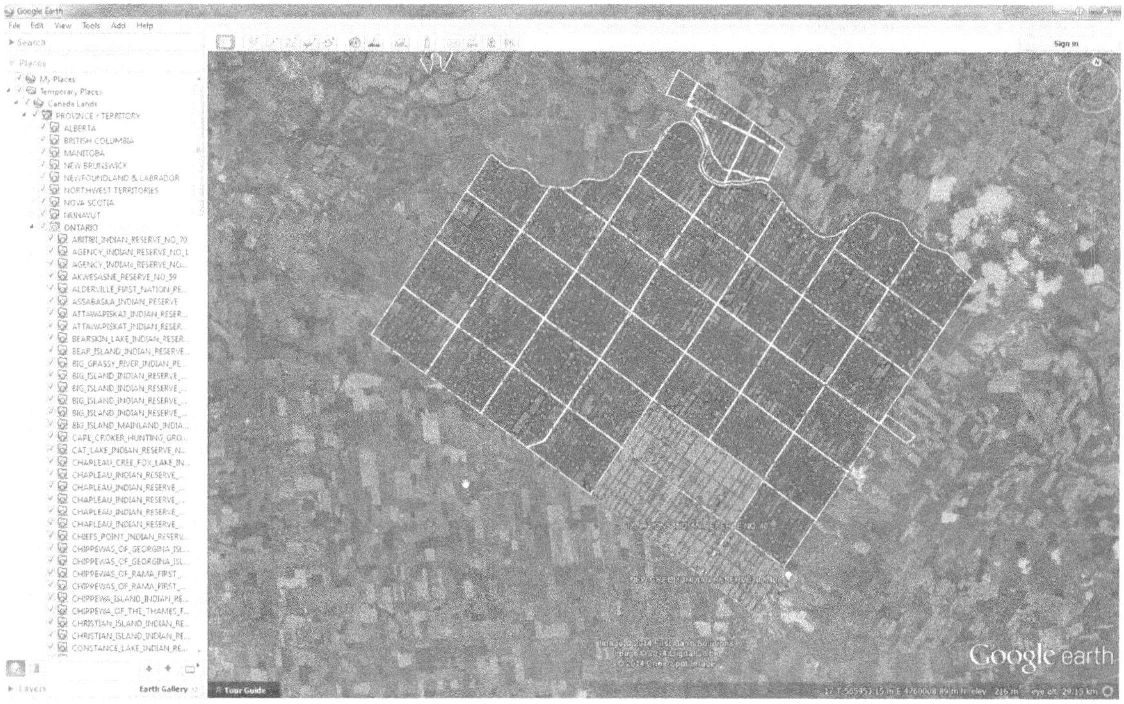

Figure 3.2. Canada Lands Overlay in Google Earth.

Source: Google Earth 7.1. Map data: 2014 First Base Solutions, 2014 Digital Globe; Canada Lands Survey System, available from http://clss.nrcan.gc.ca/googledata-donneesgoogle-eng.php

In her course, Professor Sullivan introduces students to the role of spatial relationships in archaeology especially as they relate to aspects of material culture. Using the Ancient Egyptian site of Amarna located along the Nile River, Dr. Sullivan provided students with an Excel spreadsheet of artifacts found over many decades of excavations at the site. Included in the spreadsheet is a description of the item as well as material type, but also most importantly, the location of where the artifact was found is also included. In fact, quite a bit of detail is given for the location of each artifact including the area/suburb of the ancient city, subdivision/building label, which room inside the building, and stratigraphic details detailing how deep in the ground the artifact was found. To create your own Google Earth assignment, you can find the spreadsheet at the Amarna Project website: <http://www.amarnaproject.com/pages/recent_projects/material_culture/small_finds/database.shtml>.

Using this spreadsheet, Professor Sullivan had her students select many of the artifacts and geo locate them using map overlays of the ancient city that were provided and adding them into Google Maps and Google Earth. Once mapped, students were soon able to identify the areas where particular artifact types were clustered, and patterns and potential relationships emerged, leading to research questions. For example, based on the artifact types, it could be argued that particular buildings could be homes, administrative centers, or recreational facilities.

The Amarna dataset is so rich, Professor Sullivan also crafted a second Google assignment in which students focus on a particular building excavated at the site. Students recreate the building in 3D using Sketchup and then import the virtual building into Google Earth for viewing. As part of their assignments, students needed to pay attention to the placement of the building and its functions as determined by its location in the ancient city (Sullivan, 2011).

Teaching Historical Studies in Google Earth

Digital Humanities is an ever growing area in academic research and learning, and Google Earth is one of the most frequently cited tools for helping students visualize aspects of Humanities research and unlock new knowledge. One of the most interesting teaching models of this was offered in an upper level course at the University of California Merced in 2010, which involved students learning about the "Silk Roads" that connected Europe to the trading routes to Asia and Africa. In their course assignment, students were asked to choose a famous "Silk route" from their course readings and prepare a research paper about the history of the route, famous travelers who had taken it, and a description of the goods that commuted along their chosen "Silk route." Ruth Mostern, the course instructor, also asked students to "mine" the route for spatial data, leading the students to create gazetteers of places the traveler visited, and to map the entire length route, including short descriptions of what it must have been like for the traveler at particular stops along the way. As part of their final submission, the students included Google Earth map mashups that featured their research papers and all the supplementary geographic material covering the traveler's narrative. Mostern even included a Google Earth video of the students' work on her website (Mostern and Gainor, 2013).

Teaching Language Learning using Google Earth

As with many software applications that have global appeal, Google Earth can be deployed in a wide variety of language settings. This can make Google Earth a useful tool for students who are learning a second language. The ease of using Google Earth plus its plethora of embedded resources such as Wikipedia (also available in many languages), makes it ideal as an experiential learning activity in a classroom. As presented at a conference in 2013, educators at the University of Toronto had an opportunity to do this in a third year "French as a Second Language" (FSL) course, which involved students learning how to use Google Earth in French as part of their term project (Brown, Hong, and Nicholson, 2013). Their presentation can be found at <http://www.accessola2.com/superconference2013/sessions/1303.pdf>.

To begin, students first had to learn the basic functionality of Google Earth. Following the changing of the Google Earth language settings from English to Français (see table 3.3), students performed a scavenger hunt of Quebec City and learned more about the city, while at the same time learning about the Google Earth interface and its French terminology.

After becoming adept at using Google Earth, students then worked in pairs to research a country that is a member of "la Francophonie," an international organization devoted to promoting French language and culture <http://www.francophonie.org/>. Students then produced a Google Earth mashup and animated tour of that country which they then presented to the instructor and class in French. The students were evaluated not just on their Google Earth skills, but also on their research skills, storyboarding, and presentation in the French language.

"Teaching the History of Psychology" in Google Earth

Google Earth is of course most often associated with land-based research and teaching, but as discussed throughout this chapter, Google Earth can also be adapted for teaching in other disciplines. One of the most surprising examples is discussed in an issue of

Table 3.3. Google Earth for Language Learning: Terminology from English to French

ENGLISH	FRENCH
File	Fichier
Edit	édition
View	Affichage
Tools	Outils
Add	Ajouter
Help	Aide
Sign In	Se connecter
Search	Rechercher
Get Directions	Itinéraire
History	Historique
Places	Lieux
My Places	Mes lieux préférs
Temporary Places	Lieux temporaires
Layers	Calques
Earth Gallery	Galerie Google Earth
Primary Database	Base de données primaire
Borders and Labels	Frontieres et légends
Photos	Photos
Roads	Routes
3D Buildings	Bâtiments 3D
Ocean	Océan
Weather	Météo
Gallery	Galerie
Global Awareness	Sensibilisation mondiale
More	Plus
Hide Sidebar	Masquer la barre latérale
Add Placemark	Ajouter un repére
Add polygon	Ajouter un polygone
Add path	Ajouter un trajet
Add Image overlay	Ajouter une superposition d'image
Record a tour	Enregistrer la visite
Show historical imagery, Use the time slider to move between acquisition dates	Afficher des images anciennes. Pour changer la date, faites glisser le curseur

(continued)

Table 3.3. *(Continued)*

ENGLISH	FRENCH
Show sunlight across the landscape. Use the time slider to set the time of day	Lumiére du soleil en function de l'heure du joir: choisir une heure a l'aide du curseur chronologique
Switch between Earth, sky, and other planets	Passer de la Terre au ciel ou a une autre planéte
Show Ruler	Afficher la régle
Email	Envoyer par e-mail
Print	Imprimer
View in Google Maps	Afficher dans Google Maps
Adjust Opacity	Ajuster l'opacité
Full Screen (F11)	Plein écran

Psychological Reports, in which instructors from Oklahoma State University use Google Earth in their "History of Psychology" course. Noting that history of psychology texts rarely include maps of historically important sites or even discuss how geography can influence the history and spread of ideas, the instructors sought to improve the engagement of students in the course material by illustrating lectures with Google Earth Tours of famous sites from the history of psychology such as the University of Leipzig and Sigmund Freud's home in London. In general, the authors found that the students were "favorable" to the use of Google Earth helping them visualize those places as they are seen today. This engagement carried over into several student projects supervised by the instructors. Groups of students were asked to select a historical movement in psychology such as Structuralism, Functionalist, and Behaviourist, and develop Google Earth "tours" of each "movement" including locations of universities, meeting places, and homes of the prominent psychology figures involved. Also included were embedded photos, text, and other information related to each movement (Stevison, Biggs, and Abramson, 2010).

Google Earth and Earth Science: Opportunities for Teaching and Learning Assessment

As Google Earth is a virtual representation of our planet, it is most often used as a teaching tool in Geography and other Earth Science related courses. In recent years, many university educators from those areas have been looking at effective teaching and learning models that can utilize Google Earth and at the same time properly assess whether students are in fact learning effectively.

Blended Learning

As discussed in their online text, *Pedagogy of Climate Change*, Professor Haslett and his colleagues at the University of Wales see Google Earth as an ideal blending learning tool for advancing students' understanding of climate change (Haslett et al., 2011). Blended learning has been defined as a mixed combination of self-learning using the Internet and digital media with the occasional physical presence of an instructor. In one of the chapters, Haslett and his colleagues discuss a blended learning activity in which students use published paleoclimate data collected from deep sea cores off the coast of Mauritania, to assess aridity in the Saharan desert around the capital city of Nouakchott, and how such a phenomenon is impacting the city's continued urban development. The students then compared the results they found with the course lecture material previously given and assigned readings, which discuss overall climate change in Mauritania and its impact on other African nations along the Atlantic coast.

Following the assignment, the students were asked to fill out a survey to get their feedback on using Google Earth for this climate change exercise. Of particular note for Haslett and his colleagues was the overwhelming response to Google Earth by students in the class. When asked if Google Earth was useful, all 32 students agreed that it was. When asked if it improved their awareness and enhanced their understanding of geography, only a couple of students indicated that Google Earth had not done so. Moreover, almost all of the students agreed that they would find Google Earth to be a useful tool for learning more about global environmental issues and sustainable development (Haslett et al., 2011).

Making Google Earth the focus of "student-centered, inquiry-based" learning is one of the most noticeable differences between use of Google Earth in K-12 assignments and in post-secondary courses. Rather than have an instructor at the front of the room demonstrating Google Earth for students or having students just look up information using the application and answering questions on a worksheet, many post-secondary instructors provide "scaffold-like" assignments that direct students into Google Earth and allow them to create their own placemarks, polygons, and add their own geotagged photos and maps to help answer research questions, or even better, create a tour which can serve as a story or as animated presentation of the research. As noted by many educators, Google Earth offers tremendous opportunities to break away from traditional teaching styles and involve their learners in inquiry-based learning based practises (Johnson, Lang, and Zophey, 2011).

A Bloom's Taxonomy Approach to Google Earth

Educators at Mercyhurst College in Pennsylvania were especially innovative in creating a Google Earth assignment for their students in which learning assessment went beyond just grades, but could also be matched to the Bloom's Taxonomy, which includes six levels of thinking (see figure 3.3).

These six levels have learners start at the bottom with "Knowledge" and then move upwards in terms of intellectual challenge toward "Comprehension," and passing through "Application," "Analysis," "Synthesis," and reaching the summit of "Evaluation," which is defined as the most difficult domain of learning.

In the assignment designed at Mercyhurst, students looked at earthquake and volcanic activity in the Pacific Ocean and were asked in stages to demonstrate their knowledge by inserting placemarks to indicate three subduction zones after watching a class video on the topic. "Comprehension" was tested by students exploring ridges and subduction zones on the ocean floor in Google Earth, and then answering questions posed by the instructor using clickers. "Application" comes from a writing component where students are asked to write a set of characteristics of subduction zones based on their course readings, and explorations in Google Earth. "Analysis" includes students inferring the location of volcanos near subduction zones when mapped in Google Earth. "Synthesis" starts to bring it all together with hypothesis formulation as to why volcanos form where they do and

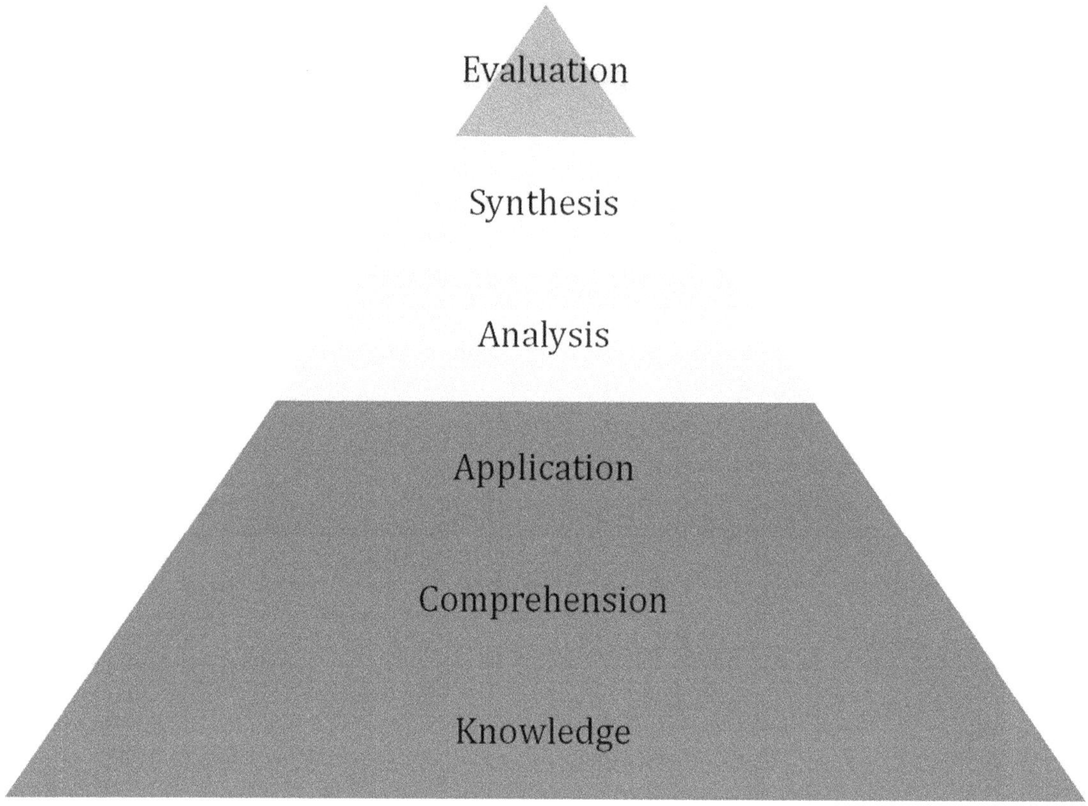

Figure 3.3. Bloom's Taxonomy Pyramid.

is presented in a five-stop Google Earth tour. Evaluation completes the learning process for the students when they have a chance to evaluate their hypothesis by comparing their "tour" with other students and creating a table listing what does and does not support their hypothesis (Johnson, Lang, and Zophey, 2011).

"Scaffolding" Assignments for Google Earth

Following a Bloom's Taxonomy approach for introducing Google Earth into the classroom can be a highly effective way for setting up a "scaffolding" approach to the mapping technology. "Scaffolding" is a means to help students select the appropriate tools and resources to answer a research question or resolve a problem. This can involve the use of modeling or a step-by-step approach that students can follow to build their knowledge base around a technology. With every new step and new level of knowledge attained, students develop confidence and new skills they can put toward their research. Scaffolding the use of Google Earth into assignments in fact is considered by many educators to be an essential step for the effective use of Google Earth in post-secondary assignments, especially for students who are not majoring in an Earth Science subject, or have never taken a Geography or Geology course before. As tested at Worcester Polytechnic Institute, educators found overall learning gains when students completed structured Google Earth labs on the geography and geology of Iceland and a slightly more challenging lab on the Tonga region of the west Pacific. In their paper, the educators were keen to point out that both male and female students saw learning gains from using Google Earth when results are compared to the pre-tests which were completed prior to the Iceland and Tonga labs. Students both new to geoscience courses and those who are majoring in the subject excelled. From such assessments, post-secondary educators are encouraged to properly scaffold their Google Earth assignments to effectively match course learning objectives (Gobert, Wild, and Rossi, 2012).

Google Earth as a Place-Based Education Tool

Along with free availability and low learning curve, Google Earth has been noted as an ideal "place based" learning tool by educators. With its imagery archives containing both aerial and Street View imagery, past and present, as well as embedded layers of geo-based information covering a plethora of human activities and physical science, Google Earth has now taken a central role in "place based" teaching strategies in many higher education courses. One such course is in Physical Geology taught at the California State University-Chico where the instructors link the concept of "sense of place" to learning geological concepts and phenomena. In their experiences prior to utilizing Google Earth in the classroom, Professor Julie Monet and Todd Greene found students had difficulty grasping both geological concepts and the "big picture" relationship of geological events. With so much geo-tectonic activity in their local area in northern California, including the San Andreas Fault, the instructors believed that they could develop a "sense of place" in their students' learning through the use of satellite images and Google Earth. By understanding their local area better with these tools, students would also better understand the geological phenomena happening around them. In their paper, which was published in the *Journal of Geoscience Education*, Monet and Greene adopt a scaffolding approach for bringing Google Earth into the course. They outline their activities including having students exposed to Google Earth from the very first week of class including "live" demos and in-class exercises which allowed them to become proficient users of the application. The instructors were especially pleased that students were soon able to answer assigned questions and pose questions of their own to the faculty by sending their own custom KMZ files by e-mail.

With students quickly gaining confidence in using Google Earth, the instructors set about creating a semester long project using a satellite image of northern California and Google Earth which would accomplish the two main objectives of the course. These include:

1. improving students' understanding of geological concepts introduced in class and its connection to their local place
2. developing a "scaffold" for students to make connections between the different geological concepts as part of a bigger picture

Working in small groups, the students were introduced to a geologic topic as well as an assigned area with a Google Earth KMZ file. A few of the geological topics assigned in the course included: Sedimentary environments, Structural geology, Plate tectonics, Earthquakes, and Groundwater. Students then had to answer a series of questions about the topic and its impact on the assigned part of northern California. As part of their final course deliverables, students submitted a Google Earth Overlay file, a PowerPoint Digital "Concept Map" poster, and a ten-minute presentation to the class utilizing each of these tools to illustrate what they had learned about their geologic topic and their broader understanding of geological happenings in their local place. In many instances, it was discovered that the students had visited those particular areas that were highlighted for each geological topic, providing an additional interest and a deeper understanding of the geological concept (Monet and Greene, 2012).

Google Earth as an "eportfolio": A Deep Learning Approach to Spatial Literacy

As discussed throughout the chapter, Google Earth offers many possibilities as a teaching and learning tool. The ability to add and geolocate just about any type of information, and then visualize it complete with audio files and/or save it as a video makes Google Earth a valuable presentation and analysis tool for students from many disciplines. Nevertheless in looking at such assignments, a scaffolding approach seems to be the most effective way to make all students comfortable with learning in a Google Earth environment. Postsecondary educators at Penn State suggest, in fact, that Google Earth makes an excellent "eportfolio" for students looking for ways to collect, access, and display research for a course, particularly in Geography.

Shocked by the 2006 Roper-National Geographic survey that found many American students could not locate Iraq on a map or identify the state of Louisiana after Hurricane Katrina, the Penn State educators began to look at Google Earth as means to lift the spatial literacy of their students, especially in courses such as "Landforms of the World," "Water: Science and Society," and an Earth Science Seminar on "Fire." In each, students are asked to develop a Google Earth eportfolio, beginning with a simple data collection using a GPS device from which they then mapped as placemarks. Over the length of the course, the students populate their eportfolio with more data, audio and video clips, podcasts, as well as other information related to their chosen or assigned topic. Once complete, their eportfolios can be shared with fellow students and the instructors.

To get started with Google Earth eportfolios, the Penn State educators suggest:

1. Start with a "Show and Tell" demo of Google Earth highlighting different components of the interface, the embedded layers, and the search functionality.
2. Provide computers for the students so they can navigate to a location and then create placemarks, edit them, and even learn some basic HTML coding which will help them format the information.
3. Demonstrate the embedding of images in Google Earth, with an emphasis on copyright-free, public domain imagery such as those available through the Creative Commons search on Flickr and Wikipedia Commons.
4. Discuss the recording and/or embedding of audio files and podcasts in Google Earth.
5. Provide students with a practice assignment with a short turnaround time so they can quickly get comfortable with using Google Earth.
6. Allow students to submit a rough draft of their assignment halfway through the term to motivate them to start early on the project and also to discover any technical issues they have with embedding links and saving ALL of their data as intended.

To provide an effective assessment of the Google Earth eportfolios, the instructors also created marking rubrics for each course that assessed not just the course content covered, but also how much the students put toward "completeness" of the project, how "open" they were in presenting and weighing evidence versus speculation, and how deep they reflected on their topic. Students were also marked on the Operational Fundamentals of Google Earth, such as checking that all their links work and that their media displays as planned, as well as on the Organizational Structure of their Google Earth eportfolio, including being clear and consistent (Guertin et al., 2012).

Key Points

Google Earth offers many instruction possibilities for librarians wishing to engage students, and even teachers and faculty. The examples described in this chapter highlight what is being taught every day in classrooms ranging from K-12 to university and college level curriculum. Each has their own learning curve from the basic use of Google Earth as a reference resource, with students looking up coordinates and place names and checking them off a list, to having students create their own Google Earth mashups and "e-portfolios" of new research.

All in all, Google Earth is providing a dynamic and interactive educational experience to the classrooms which will open up all kinds of opportunities for experiential learning for students. Many studies and real-life examples have demonstrated that Google Earth is an effective platform for teachers and students to view, understand, and analyze the Earth across a wide variety of subjects. It not only supports spatial thinking but also helps develop critical analytical skills, while preparing students to use GIS in future work. Librarians and other educators are encouraged to follow some of the examples provided in this chapter and embed Google Earth technology into the course curriculum enabling students to go beyond traditional learning methods to discover the world around them in a unique and interactive way.

- K-12 courses are using Google Earth in a variety of subjects to learn about the world around them
- Course instructors are finding positive results in using Google Earth as a model for spatial representation and for gaining spatial literacy skills
- Google Earth is being utilized in a broad array of disciplines across university and college curricula
- Post-secondary instructors and librarians are using Google Earth to assess learning outcomes, and recommend a "scaffolding" approach to help students maximize their use of the application in support of learning course content.

The next chapter will focus on advanced functionality of Google Earth, providing readers will the skills necessary to turn Google Earth into a map making and spatial analysis program. Users will learn how to create customized KML files, and how to use the product as a GIS program.

References

Almquist, Heather, Lisa Blank, and Jennifer Estrada. 2012. "Developing a Scope and Sequence for Using Google Earth in the Middle School Earth Science Classroom." *The Geological Society of America Special Paper* 492: 403–412.

Andrews, Geoffrey. 2006. "Laptops + Challenging Curriculum = Student Success." *Techniques: Connecting Education & Careers* 81, no. 2: 42–44.

Bodzin, Alec. 2008. "Integrating Instructional Technologies in a Local Watershed Investigation with Urban Elementary Learners." *The Journal of Environmental Education* 39, no. 2: 47–56.

Bodzin, Alec, David Anastasio, and Violet Kulo. 2014. "Designing Google Earth Activities for Learning Earth and Environmental Science." In *Teaching Science and Investigating Environmental Issues with Geospatial Technology: Designing Effective Professional Development for Teachers*, edited by James MaKinster, Nancy Trautmann, and Michael Barnett, 296–321. Dordrecht, Netherlands: Springer.

Britt, Judy, and Gus LaFontaine. 2009. "Google Earth: A Virtual Globe for Elementary Geography." *Social Studies and the Young Learner* 21, no. 4: 20–23.

Brown, Robert, Rosa Hong, and Andrew Nicholson. 2013. *Mapping Your Way to Second Language Teaching & Learning.* Presentation at the Ontario Library Association Superconference, Toronto, Ontario; February 1, 2013. http://www.accessola2.com/superconference2013/sessions/1303.pdf.

Butler, Douglas. 2008. "Having Fun with Google Earth." *Mathematics Teaching Incorporating Micromath* 209: 28–30.

Castek, Jill, and Jessica Mangelson. 2008. "Thinking Outside the Book: Reading the World with Google Earth." *Book Links* 17, no. 5 (May): 10–11.

Doering, Aaron, and George Veletsianos. 2008. "An Investigation of the Use of Real-Time, Authentic Geospatial Data in the K–12 Classroom." *Journal of Geography* 106: 217–225.

Geography Education Standards Project. 1994. *Geography for Life: National Geography Standards.* Washington, DC: National Geographic Society.

Gobert, Janice, Steven C. Wild, and Lisa Rossi. 2012. "Testing the Effects of Prior Coursework and Gender on Geoscience Learning with Google Earth." *Geological Society of America Special Paper* 492: 453–468.

Guertin, Laura, Christopher Stubbs, Christopher Millet, Tsan-Kuang Lee, and Matthew Bodak. 2012. "Enhancing Geographic and Digital Literacy with a Student-Generated Course Portfolio in Google Earth" *Journal of College Science Teaching* 42, no. 2 (November): 32–37.

Guth, Peter. 2012. "Automated Export of GIS Maps to Google Earth: Tool for Research and Teaching." *The Geological Society of America Special Paper* 492: 165–182.

Haslett, Simon K. 2009. "Prior Use of Google Earth by Undergraduate Geography Students." *Planet* 22: 43–47.

Haslett, Simon K., Andrew Skellern, Matthew Chilcott, and David Longman. 2011. "Climate Change Education Through a Blended Learning Google Earth Exercise." In *Pedagogy of Climate Change*, edited by Simon K. Haslett, Derek France, and Sharon Gedye, 112–127. Plymouth, United Kingdom: Higher Education Academy, Geography, Earth and Environmental Sciences.

Jenny, Geraldine. 2011. "Taking Student Learning Out of This World: Integrating Google Earth in the Elementary Classroom." *National Social Science Technology Journal* 1, no. 2. http://www.nssa.us/tech_journal/volume_1-2/vol1-2_article5.htm.

Johnson, Nichols D., Nicolas P. Lang, and Kelley T. Zophey. 2011. "Overcoming Assessment Problems in Google Earth-based Assignments" *Journal of Geoscience Education* 59: 99–105.

Misegadis, Erin. 2011. "Flat Stanley and Google Earth." *Discovery Education Blog.* October 9. http://blog.discoveryeducation.com/blog/2011/10/09/flat-stanley-and-google-earth/.

Monet, Julie, and Todd Greene. 2012. "Using Google Earth and Satelliate Imagery to Foster Place-Based Teaching in an Introductory Physical Geology Course." *Journal of Geoscience Education* 60: 10–20.

Mostern, Ruth, and Elana Gainor. 2013. "Travelling the Silk Road on a Virtual Globe: Pedagogy, Technology, and Evaluation for Spatial History." *Digital Humanities Quarterly* 7, no. 2. http://www.digitalhumanities.org/dhq/vol/7/2/000116/000116.html.

National Research Council. 2006. *Learning to Think Spatially.* Washington, DC: National Academies Press.

Rakshit, Rahul, and Yelena Ogneva-Himmelberger. 2008. "Application of Virtual Globes in Education." *Geography Compass* 2, no. 6: 1995–2010.

Stevison, Blake K., Patrick T. Biggs, and Charles I. Abramson. 2010. "Google Earth as a Source of Ancillary Material in a History of Psychology Class." *Psychological Reports* 106, no. 3: 665–670.

Sullivan, Elaine. 2011. "The Old (World) and New (Technologies): Teaching Archaeology in the Age of Google." *Near Eastern Archaeology,* April 12. http://www.asor.org/pubs/nea/The%20Old%20%28World%29%20and%20New%20%28Technologies%29.pdf.

Whittingham, Jeff, Stephanie Huffman, Wendy Rickman, and Cheryl Wiedmaier, eds. 2013. *Technological Tools for the Literacy Classroom.* New York: IGI Global.

CHAPTER 4

Google Earth for Mapping and Sharing

IN THIS CHAPTER

- ▷ Getting "under the hood" of Google Earth: KML and HTML scripting
- ▷ Migrating GIS Data to KML: Online and desktop conversion tools
- ▷ Discovering advanced functions in Google Earth
- ▷ Using Google Earth Pro to its fullest potential

GOOGLE EARTH OFFERS USERS A VARIETY of mapping and visualization functions. As discussed in chapter 1, Google Earth easily allows users to mark up aerial maps with placemarks, lines, and polygons. Google Earth also has many other more advanced functions and capabilities that librarians and other instructors can take advantage of to promote teaching and learning in the classroom. In some instances these functions are part of the program, while others involve the use of external apps and other web services that have been set up to support users of Google Earth. This chapter will provide comprehensive instructions on how to access and utilize these functions and web services in support of teaching and learning in the classroom.

Getting "under the hood" of Google Earth: KML and HTML Scripting

As discussed in chapter 1, KML (Keyhole Markup Language) is the primary code used in the making and displaying of placemarks, polygons, paths, and other files that are added into a Google Earth map. Using the Google Earth interface and functions available from the menus, a user can make some interesting and content-rich visualizations. Nevertheless, much more is possible for Google Earth users who wish to work under the hood of

Google Earth; that is to modify or create their own KML code to visualize elements in Google Earth.

For users who are already comfortable with coding, such as using HTML for creating webpages, working with KML code is quite similar. As can be done with writing HTML code, KML code can be created using simple text applications like Notepad or Textedit. With KML code, users can add some additional features and animations to their map such as time sliders, legends, and further customizations to placemark "balloons."

To see what a raw KML file looks like ("looking under the hood"), simply take the KML file of interest and open it using any text application such as Wordpad on a Windows computer or Textedit on a Mac. Figure 4.1 is an example of raw KML code. This file in fact shows the distribution of trees along Manhattan streets and comes from a Street Tree Census, made available through the New York City Open Data website <https://data.cityofnewyork.us/Environment/Street-Tree-Census-Manhattan-/e6n3-m3vc>. This is a typical example of a KML file available for download to the general public.

KML Interactive Sampler

As can be seen with webpages or HMTL files, there can be a lot of code associated with KML files. Even a relatively simple KML file can mean lines and lines of code which

```
Street Tree Census (Manhattan).kml
<?xml version="1.0" encoding="UTF-8"?><kml xmlns="http://www.opengis.net/kml/2.2" xmlns:xsi="http://www.w3.org/
2001/XMLSchema-instance" xsi:schemaLocation="http://www.opengis.net/kml/2.2 http://schemas.opengis.net/kml/2.2.0/
ogckml22.xsd">
   <Document xmlns:atom="http://purl.org/atom/ns#">
      <name>geo_e6n3-m3vc:geo_e6n3-m3vc-1</name>
      <LookAt>
         <longitude>-73.96454870101272</longitude>
         <latitude>40.787527441422185</latitude>
         <altitude>25994.95337399661</altitude>
         <range>21005.285804913452</range>
         <tilt>0.0</tilt>
         <heading>0.0</heading>
         <altitudeMode>clampToGround</altitudeMode>
      </LookAt>
      <Placemark id="geo_e6n3-m3vc-1.1">
         <name><![CDATA[geo_e6n3-m3vc-1.1]]></name>
         <description><![CDATA[<h4>geo_e6n3-m3vc-1</h4>

<ul class="textattributes">

  <li><strong><span class="atr-name">OBJECTID_1</span>:</strong> <span class="atr-value">16</span></li>
  <li><strong><span class="atr-name">TREEID</span>:</strong> <span class="atr-value">186</span></li>
  <li><strong><span class="atr-name">ONSTREET</span>:</strong> <span class="atr-value">TRINITY PLACE</span></li>
  <li><strong><span class="atr-name">PARITY</span>:</strong> <span class="atr-value">odd</span></li>
  <li><strong><span class="atr-name">CROSSSTREE</span>:</strong> <span class="atr-value">THAMES STREET</span></
li>
  <li><strong><span class="atr-name">CROSSSTR_1</span>:</strong> <span class="atr-value">RECTOR PLACE</span></li>
  <li><strong><span class="atr-name">BUILDINGNU</span>:</strong> <span class="atr-value">74A</span></li>
  <li><strong><span class="atr-name">BUILDINGST</span>:</strong> <span class="atr-value">TRINITY PLACE</span></
li>
  <li><strong><span class="atr-name">TREELOCATI</span>:</strong> <span class="atr-value">1.0</span></li>
  <li><strong><span class="atr-name">SITE</span>:</strong> <span class="atr-value">4</span></li>
  <li><strong><span class="atr-name">TREEPIT</span>:</strong> <span class="atr-value">1</span></li>
  <li><strong><span class="atr-name">TREECONDIT</span>:</strong> <span class="atr-value">1.0</span></li>
  <li><strong><span class="atr-name">DIAMETER</span>:</strong> <span class="atr-value">6.0</span></li>
  <li><strong><span class="atr-name">SPECIES</span>:</strong> <span class="atr-value">PYCA</span></li>
  <li><strong><span class="atr-name">BOROUGH</span>:</strong> <span class="atr-value">Manhattan</span></li>
  <li><strong><span class="atr-name">ZIPCODE</span>:</strong> <span class="atr-value">10006</span></li>
  <li><strong><span class="atr-name">FID_1</span>:</strong> <span class="atr-value">0</span></li>
  <li><strong><span class="atr-name">COMMDIST</span>:</strong> <span class="atr-value">101</span></li>
  <li><strong><span class="atr-name">_SocrataID</span>:</strong> <span class="atr-value">3gvy-8s2m</span></li>
</ul>
]]></description>
         <LookAt>
            <longitude>-74.0125844877534</longitude>
            <latitude>40.708353445054875</latitude>
            <altitude>0.0</altitude>
            <range>0.0</range>
            <tilt>0.0</tilt>
            <heading>0.0</heading>
```

Figure 4.1. Street Tree Census (Manhattan) in raw KML format.

need to be accurately inputted and saved for it to be displayed properly in Google Earth. Fortunately, with KML being recognized as an open standard by the OGC (as discussed in chapter 2), Google does make it much easier for users to modify and adapt KML files. Available online is the KML Interactive Sampler <https://kml-samples.googlecode.com/svn/trunk/interactive/index.html?hl=es>, which provides KML code to Google Earth users who are looking to create particular functions or effects in their Google Earth map.

The Sampler contains over forty KML scripts that users can take and modify or copy and paste into an existing KML file. What especially endorses the KML Interactive Sampler is its ease of use. Upon opening the website, users will see three windows. On the left, there are specialized KML script options. The middle window displays the KML script itself, row by row. The rows are also numbered to make reading and deciphering the code that much easier for users. On the right side of the screen is a Google Earth window, which requires a Google Earth plug-in to be installed in the Internet browser for it to display. This window allows the user to visualize the KML script or special features as it would appear in Google Earth. Although the Google Earth viewer offers a nice complement to using the sampler, it is by no means necessary for someone to take advantage of the scripts in the KML Interactive Sampler. The Google Earth Plug-in works better in some browsers than in others, and specifically on a Mac it works fine in Safari, but was unable to load in Firefox. For Windows PCs, it runs well in Google Chrome and Firefox.

Returning to the left side of the screen, there are several dozen KML sample scripts available. As outlined below in table 4.1: KML Interactive Sampler Choices, these are broken up into five sections: "Placemarks and Geometries," "Other Features," "Other Basic Topics," "Advanced Topics," and "Google's KML Extensions." In each section there are subcategories that include specialized KML scripts for "Placemarks," "Lines and Paths," "Polygons," "Models," and many more. Table 4.1 includes all of the KML Interactive Sampler Script options that can be extracted and used with personal KML Files.

The KML Interactive Sampler also provides links to further samples that have been created and shared on a KML-samples wiki page. For example, users can download KML code that will help create "Mouseover" effects over Google Earth features, or make them extrude or pop up from the map. Users can also add their own scale bar as an overlay to the Google Earth map. Many KML samples are available and can also be added to users' own wiki if they are signed in using a Google account.

Users who wish to learn more about KML Languages can also refer back to the Google Developers pages, which have extensive reference information on using and developing KML code. For example, users will find a KML elements "class tree" on the KML Reference webpage <https://developers.google.com/kml/documentation/kmlreference?csw=1> which is interactive and provides yet more sample code for users who require their Google Earth maps to perform particular functions.

Creating Animations or Time-Sensitive Data

One of the best aspects of KML coding is being able to create animations in the Google Earth map, including setting up time displays of data to showcase changes of routes or landscapes over time. Available in the KML Interactive Sampler are KML codes for "TimeSpan," which allows users to look at different overlays over a period of time using the Time Slider. Users can download this sample of KML code and then customize it to show changes in landscapes over time. For example, a series of satellite images over Las Vegas from the 1950s to 2014 can be overlaid in Google Earth and set up as a "Time

Table 4.1. KML Interactive Sampler Choices

KML Code is available for copying and pasting for each of these Google Earth elements

PLACEMARKS AND GEOMETRIES	OTHER FEATURES	OTHER BASIC TOPICS
Placemarks (Points)	**Ground Overlays**	**Balloons**
Descriptive HTML	Large Ground Overlay	Background Color
Extruded	**Screen Overlays**	Display Modes
Floating	Absolute TopLeft	Simple
Simple	Crosshairs	**Styles**
Lines and Paths	Dynamic Right	Inline
Absolute	**Network Links**	Rollover (StyleMap)
Absolute Extruded	Fly-to-View	Shared
Relative	Multiple Instances	Style Merging
Relative Extruded	Simple	**Sky**
Tessellated	Targets	Leo
Untessellated	Simple	**Views (Camera and LookAt)**
Linear Ring	With LookAt	Ground Level
Polygons	**Advanced Topics**	Looking Down
Absolute	**Extended Data**	Looking Up
Absolute Extruded	Balloon Substitution	Roll Left
Extruded	**Regions**	Tilt
Google Campus	Polygon Fade	**Google's KML Extensions**
Relative	Scale	**Touring**
Relative Extruded	Swap Polygons (Fade)	Complete Touring Demo
Models	Swap Polygons (Pop)	Bounce flyToMode
Macky Auditorium (ResourceMap)	**Time**	Smooth flyToMode
Shared Textures	Time Spans	Pause Tour Primitive
Multi-Geometries	Time Stamps	Wait Tour Primitive
Line Strings	Photo Overlays	Animated Updates
Rollover	Seattle Space Needle	Balloon Visibility
Simple		Sound Cues
		Ocean
		Polygon w/ clampToSeaFloor Altitude Mode
		Line String w/ relativeToSeaFloor Altitude Mode

Span" KML to highlight the urban development in one of the fastest growing cities in the United States.

Another time-related sample of KML code is "Time Stamp," which allows users to link each placemark to a time stamp. For example, to map a hiking trail, users can record waypoints along the route with a GPS device, which will also record a timestamp for each waypoint. The waypoints plus the timestamps can later be imported into Google Earth and will allow users to animate their trip as the time slider is moved forward in time.

Creating Google Earth Description Balloons/Dynamic Placemarks

When exploring the embedded layers inside Google Earth, especially in the "Global Awareness" category, users will notice that the placemarks organized by groups such as Greenpeace and the WWF (World Wildlife Federation) have very different appearances. The placemarks are formatted differently, including colorful banners, photos, and videos. These are dynamic placemarks and can be created by adding some additional code to the placemark balloon.

Although many of the dynamic placemarks found embedded in Google Earth look professionally done, it is actually quite easy for anyone using Google Earth to create their own. Users can easily highlight a project by taking their data and creating dynamic balloons and placemarks by including elements such as images (like a library logo), and customizing the text font, color, and the color of the balloon itself.

The easiest way to customize a placemark balloon is to start with downloading a balloon template from the Google Earth Outreach webpages <http://www.google.com/earth/outreach/tutorials/balloon.html>. Google provides a selection of templates from which users can choose depending on what they wish to present in the balloon.

When users download a template, they can save it to their computer, but it is more useful to open it right away in Google Earth so that they can begin to edit it. Two files are actually downloaded in this step into the "Temporary Places" folder. These are the photo placemark template and a sample placemark showing what the layout should look like. As illustrated in figure 4.2, the template is really just a couple of placeholders for new text and photos to be added.

Before starting to edit the template balloon, users may want to make additional copies depending on how many of the particular templates they wish to use in their project. To edit the template, the file will need to be selected to access the Properties menu. As seen in figure 4.3, when users open the Properties window of the template placemark, they will see the

GOOGLE EARTH BALLOON TEMPLATES AVAILABLE FOR DOWNLOAD

- Wide Photo
- Tall Photo
- Text & Photo Banner
- Text & Tall Photo
- Text & Weblinks

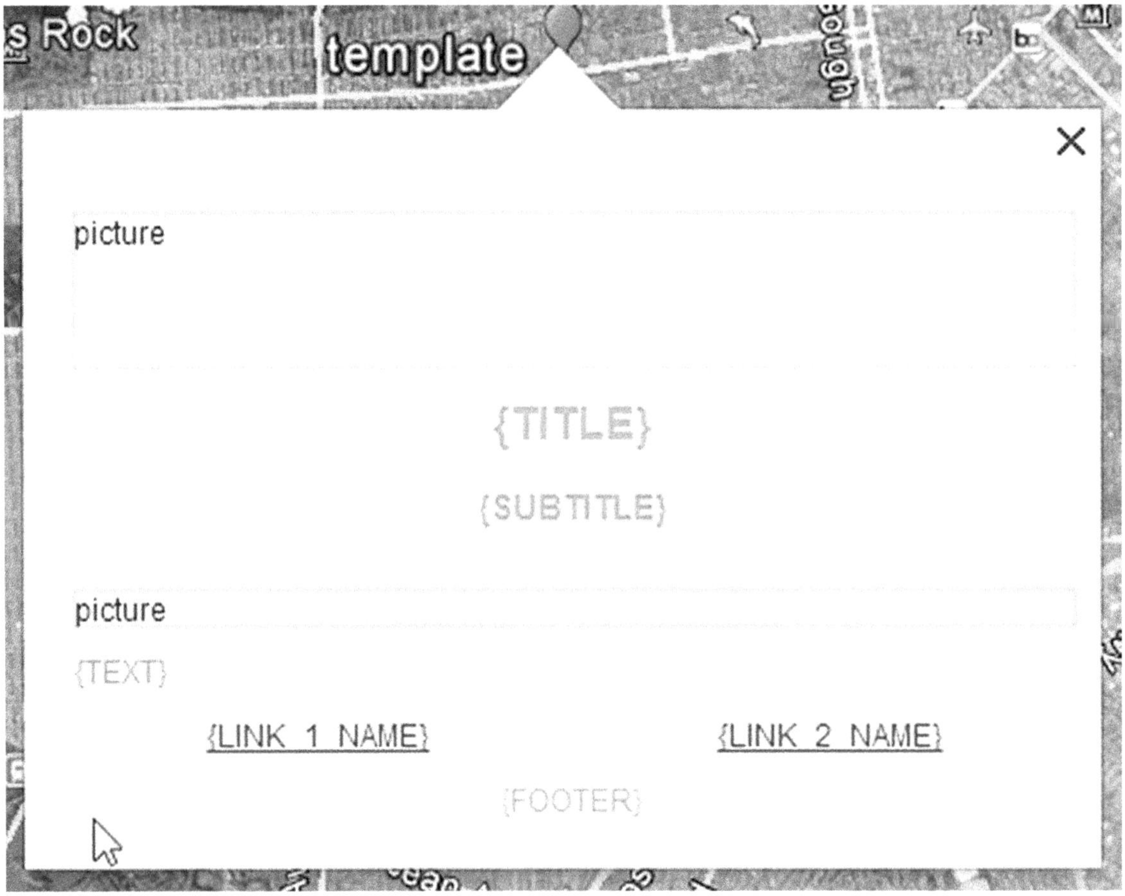

Figure 4.2. Placemark Balloon Template in Google Earth
Source: Google Earth 7.1, <http://earth.google.com/>.

HTML coding that is used to populate the balloon. Included is some text in block capital letters, which serve as a placeholder for users to enter their text and images. The placemark will need to be repositioned exactly where it should go, as well as be renamed. Clicking on "Snapshot the current view" will snap the placemark into that location.

Next, users can edit the HTML code in the Description window, a straightforward activity as the HTML can simply replace the text that is in uppercase letters with the desired placemark-specific text. The same also applies to links; users can replace the placeholder links with links to their own images and websites that are relevant for their placemark. In figure 4.4, the user has completed the adding of text, photos, and other links to finish with a professional, authoritative looking placemark.

Migrating GIS Files to KML: Online and Desktop Tools

As highlighted throughout this book, Google Earth provides powerful means for users to create their own map data and save them as KML files to share with others. Google Earth also offers a way to view KML files created by others, including maps made by fellow researchers or produced by a government agency. Oftentimes, however, a user will encounter useful map data, but they will be in a non-KML format. For example, a user may wish to map a city's growing population over several decades, highlighting neighborhoods which have seen the largest increases. On that city's open data website, only the

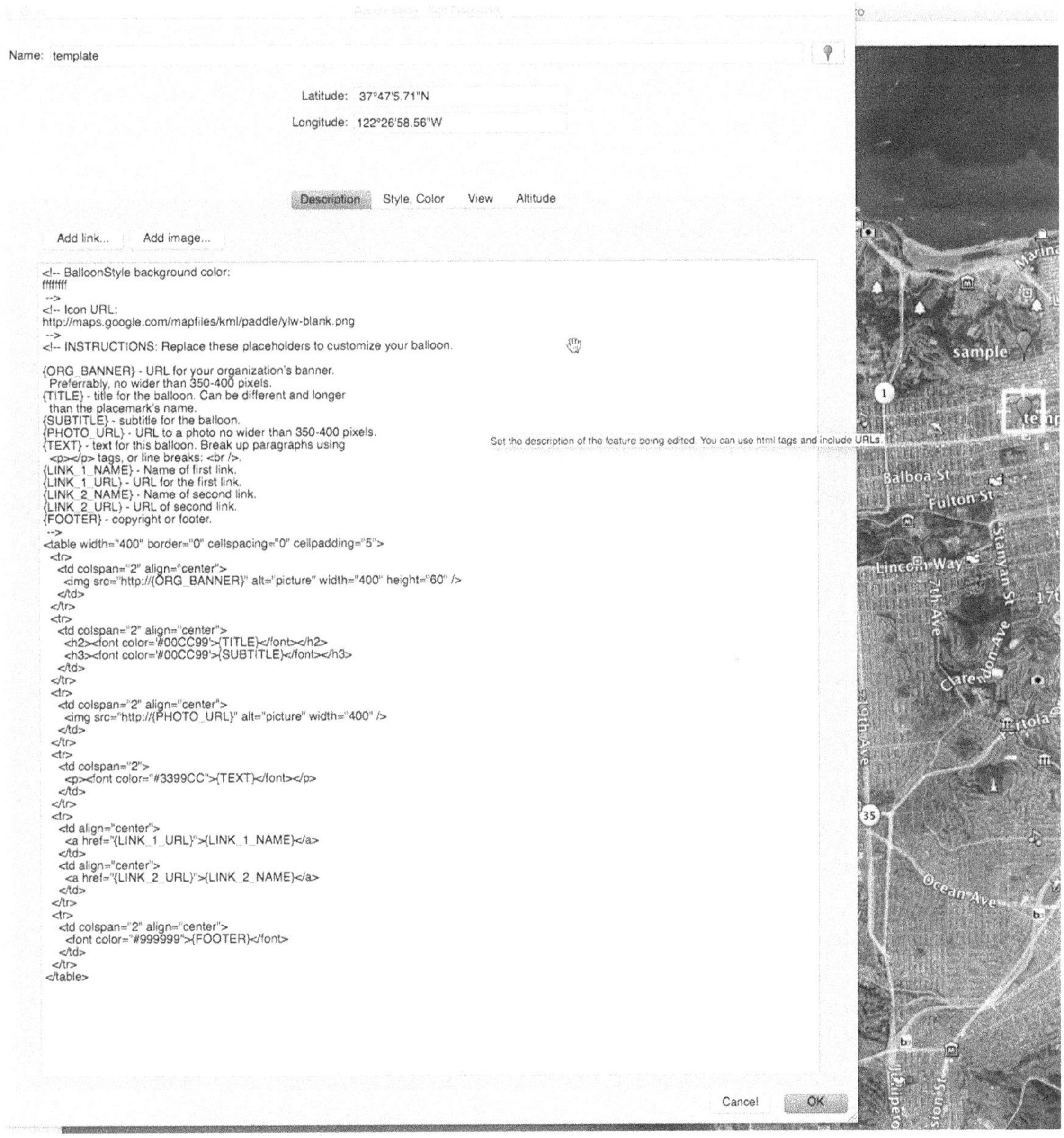

Figure 4.3. HTML coding for Balloon Template.

Source: Google Earth 7.1,<http://earth.google.com/>.

current population totals are available as a Google Earth friendly KML file, so a simple comparison of population change by comparing yearly datasets going back decades may be difficult to accomplish. When encountering these kinds of issues, it is important to remember that Google Earth is still a relatively new application and that the majority of available KML files from public agencies are focused on today's public and their interests and issues. Older and historical map files in KML continue to be rare and are often difficult to find. Materials not born digitally are often impossible to find in any GIS format let alone KML. Even with digitally born maps, it is worth remembering that many geographical information systems applications have their own file formats, some of which are proprietary and cannot be viewed in Google Earth.

Perhaps the most ubiquitous GIS file format that users will encounter is the shapefile or .shp. Shapefiles are a proprietary format produced by Esri, a privately owned GIS

Figure 4.4. Template modified for Bowdoin College Bookstore

Source: Google Earth7.1, <http://earth.google.com/>. Map data: 2015 Google. Data Source: http://www.bowdoin.edu/~disrael/google-earth/google-earth-templates/.

company based in Southern California. It is most often used in Esri's ArcGIS software, which is utilized in thousands of government offices, as well as university libraries and academic departments for research. Although there is one shapefile (.shp) for every Esri GIS vector dataset, these in fact can only work with other associated files in the same folder, including a .shx, and .dbf to provide the means for the map shapefile to work in GIS software. For best accuracy, a shapefile also should be accompanied by .prj, .sbn, and .sbx files.

In recent years Esri has relaxed the use of the shapefile standard allowing other GIS software programs to also use it, including Google Earth Pro, which will be discussed later in this chapter. Several desktop and online applications are also available which allow for the conversion of shapefiles and other GIS files into a Google Earth friendly KML format.

Shape2Earth

Shape2Earth <http://shape2earthengine.com/shape2earth/Home.html> is a free and open source website that offers a couple of tools which would be most useful for Google Earth users. One is the "Shape2Earth Engine," which is a stand-alone utility for Windows computers and allows for the converting of Esri shapefiles into a KML format. Users can download an evaluation version, which allows for a simple uploading of an Esri shapefile GIS dataset. As a user will notice with many of these GIS data converters, the shapefile needs to be in a compressed zipped format along with its associated files for a successful conversion to take place. One of the best aspects of the Engine Utility

To save it to a permanent .kmz file, right-click the tour in the Places panel, choose Save Places As…, and save the .kmz file to a local folder on the computer.

Re-recording a Tour to Add New Actions

Re-recording a tour allows users to add more actions to their tour through modifying and extending a previous tour. It requires users to re-record a new tour while playing back the old tour. To do this, the user will need to click on the Add Tour button in the toolbar to add a tour. Clicking the Record button will start capturing a new tour. After clicking the record button, the time counter starts counting and showing the duration of the new tour. The original tour can now be played by double-clicking on it in the Places panel, or selecting it with the mouse and clicking the Play Tour button underneath the Places panel. A new tour is now being recorded while playing back the original one. The original tour can be paused to add some new actions or camera movements. Once the original tour is done playing back, the recording session is still active, and therefore additional actions can be added.

Recording Narration

When creating a new tour, the user can click on the microphone button instead of the record button, which will capture audio from the computer's microphone. Clicking on the microphone button again will turn off audio recording. Audio recorded through the microphone is captured as an .mp3 file that is saved with the tour when saving as a KMZ file. When saving the Tour with audio narration, the "Save as Type" option in the Save window must be set to KMZ, not KML.

Limitations

The recorded tour can only be saved as a KMZ file and cannot be exported outside of Google Earth to be used in another program. This means that users can view the recording only in Google Earth. However, the video can be brought into Google Earth Pro and additional movie-making features are available that will allow users to save the video file as a Windows Media Video (WMV), Audio Video Interleave (AVI), Image Stream (JPG), and/or Apple QuickTime (.mov). These files can then be transferred to colleagues via e-mail or uploaded to a website. Saving to these file formats is possible due to an additional video recording feature that Google Earth Pro offers in its "Tools" menu called Movie Maker.

Working with Google Fusion Tables

Google Earth supports the importation of personal files like tables, data, and images, but they need to be in KML file format. This is why becoming familiar with file converters such as those that convert from shapefile to KML is important. Some external products, however, work really well in Google Earth as they provide an easy and seamless way to create or convert files into KML format.

One of the quickest and easiest ways to produce simple maps is to start with Google's Fusion Tables <http://www.google.com/fusiontables/Home>, which will customize tabular data to be viewed and used in Google Earth. Fusion Tables is an online data management application designed for collaboration, visualization, and publication of data.

It allows users to upload and manage map data; to map points, lines, or areas; to create pushpins; to create other types of visualizations (i.e., charts); to embed visualizations into a website; and to share and collaborate with others.

For users who wish to work with tabular data in Google Earth, it is recommended to import a table that was created in Excel into Google Fusion. The table must have location information included, such as city, address, or zip code/postal code. Once imported, the table can easily be saved as a KML file, at which point it is ready to be imported into Google Earth.

To open the file in Google Earth, the user will need to go to File, Open, and then navigate to the folder of the KML file, making sure to select Google Earth (*.kml *.kmz *.eta *.ini) from the drop-down menu next to the file name. Once imported, each row will be represented by a point on the map. Clicking on a point will open a window pop-up, showing the attribute information from the Fusion Table.

The table can easily be customized, including changing the icon on the map or a label's name. This can be done by clicking on "Properties" on the left panel. A window will appear with four tabs. The description tab contains the attribute information that previously appeared in the pop-up when the icon was clicked. The attribute information can be edited or deleted, or additional information can be added. Links and images can also be added to the window pop-up if the user desires to do so. The Style, Color tab enables users to edit the color, scale, and opacity of the label and the icon. The View tab allows a snapshot of the current view. However, users will not be able to edit the other information. The Altitude tab allows extruding of the rectangle (elevation profile) around the icon in 3D format. Choosing one of the options from the drop-down menu will set the perspective altitude, and then users can input a specific number in meters.

Using Google Earth Pro to Its Fullest Potential

As was discussed in chapter 1, Google Earth Pro is a slightly more advanced version of the regular Google Earth program. A major additional feature is its ability to import Esri shapefiles and display them like a GIS program would do, draping the features accurately over the Earth's surface. Google Earth Pro users do not need to take the extra step of converting shapefile to KML, as they can directly open the data layers in the program. Once the layers are added, however, they can easily be saved and exported out as KML files and shared with users. Currently, there are many more spatial data files available online in shapefile format than in KML file format, so downloading a copy of Google Earth Pro would be essential if you are interested in making maps and utilizing data from many sources.

As was discussed in previous chapters, GIS data is georeferenced to real-world locations on the Earth and consists of either vector data (shapefiles that are points, lines, or polygons), or raster data (i.e., aerial imagery). In order for Google Earth Pro to import the data and place it correctly on the Earth's surface, all data must be georeferenced or have a coordinate system defined. This is typically embedded into all spatial data downloaded from libraries or online. The following sections will highlight some of the more advanced functionalities of Google Earth Pro, enabling users to use sophisticated methods to map and analyze their data.

Creating and Mapping 3D Data

Shapefiles are typically accompanied with tabular data that can range from offering very little to very significant information. Typically when users import a shapefile, they will see the points, lines, or polygons displayed on the Earth, but the information that describes the geometry is not seen until the features are clicked on. What is available when clicking on the feature is dependent on the attribute table associated with the original shapefile. Some attribute information is left blank, as it may be with a shapefile that is displaying address points, and others are fairly rich, including lengths, distances, and names, as may be seen with a river file. Similar to a desktop GIS program, Google Earth Pro can classify and categorize the information available in the attribute table. A statistical shapefile that offers population density, for example, can be classified based on numerical breaks and displayed to create a colorful choropleth map. Additionally, the numbers can be extruded to showcase differences in values.

Importing a Shapefile

Shapefiles can be accessed from a number of online data portals, libraries, and websites and will need to be downloaded and saved to the local computer. One or more shapefiles can be added into Google Earth Pro by going to File, Import, and navigating to the folder where the shapefile is located. Ensure to select ESRI Shape (*.shp) from the drop-down menu and click "Import All" for the Data Import warning pop-up.

Mapping the Shapefile

Google Earth Pro has the unique feature of enabling not only mapping using shapefiles, but also mapping in three dimensions (3D). Any shapefile that is in polygon format (as opposed to point or line) has a "height" dimension that can be set to visually extrude the data. This works really well for numeric data like population. Areas that have higher population values will be presented with taller three dimensional polygons, as opposed to areas that have smaller populations which will be flat and flush with the globe.

Shapefiles can also be easily customized to display text, symbol, and color. When working with choropleth maps, a color ramp can be used to distinguish low values from the high values. These tools are available in the layer properties. The attribute information that is embedded into the shapefile can be viewed by clicking on the layer or by clicking on the polygon on the map. Shapefiles can be saved as KML or KMZ files by going to File—Save—Save Place As.

Geocoding Data

Google Earth Pro offers the additional ability to import not only vector and raster geospatial data, but also tabular files in .txt or .csv format. This is a significant feature of the upgraded version because unlike shapefiles, tabular data can be created by anyone, using their own personal data. Common tables often include GPS coordinates, like in projects that track routes or findings, or address information, often found in projects that focus on collecting a series of specific residential or business information. Taking the rows and columns of address information or GPS coordinates and displaying them as points on a map is a process called geocoding. When users import the table of addresses into Google Earth, each address is converted into a placemark that is displayed in the 3D viewer and in the Places panel. Google Earth can only ingest address information located within the United States, United Kingdom, Canada, France, Italy, Germany, and Spain.

Before a table can be imported into Google Earth Pro, it needs to be created with columns that include location data. In Excel, a Comma Delimited (.csv) file needs to be created and populated with data that includes location coordinates, such as addresses. Depending on the scale of the data (zip code/postal code vs house number), the table may need to have several columns breaking down the location information. For detailed addresses, like house numbers, there should be a column for the house number and street address, and separate columns each for the city, state/province, and zip code/postal code. Once populated and saved, the table can be opened in Google Earth Pro by going to File, Open, and navigating to the folder of the data table in .csv format, making sure to select Generic Text (.*txt .*csv) and clicking "Open."

The geocoding process will be guided by Google Earth's Data Import Wizard, a function that helps users customize their geocoded results. The Wizard starts up as soon as the table is imported. For Field Type option, select Delimited; for Delimited option, select Comma. There will be a preview of the data at the bottom of the window, depicting how the data will look once imported. If any of the addresses in the table were not formatted correctly or had errors in them, then a dialog box will announce the problem data rows.

Once the geocoding has been successful, a Style Template dialog box will appear, prompting the user to select a name, color, and icon for the table and map. Once completed, Google Earth will then display the address data as icons in the 3D viewer. Users have the option to edit the geocoded point placemarks later if they choose to. As with other placemarks, users can change the icon symbol (include a customized icon), size, or color by clicking on any of the icons and clicking Properties and Edit Placemark.

Because not all users like to prepare their data beforehand, and because many simply do not know how to use the geocoding tools found in Google Earth Pro, manually typing in addresses in the regular version of Google Earth works as well. For those wishing to collect a small number of addresses and plot them on the map, manually creating placemarks for each address point is an alternative way to geocode.

Vector Regionation

Oftentimes, users will have large amounts of map data, but may only need or want the data to display in a particular area of the Earth. For example, someone wishing to map the locations of Farmers Markets that sell organic products in the United States can do so; however, the point data or placemarks that will appear in the map will make the display too packed or "too busy" for the user to read or be useful in any way. To better display the data, the user can apply the regionate function available in the Google Earth Pro Tools section.

Vector Regionation will allow for the displaying of points based on the altitude users are at in Google Earth. The closer the users zoom in, the more points (or placemarks) are displayed. If the user is just interested in Farmers Markets in California, zooming into the state in Google Earth will just display Farmers Market locations for the state of California in the United States.

Selecting Regionate brings up a window, which allows the user to specify the "input file," which must be a text file, a CSV file, or an existing KML file. The user then specifies

CHAPTER 5

Google Earth in Practice: Self-Guided Tutorials

IN THIS CHAPTER

- ▷ Creating dynamic placemarks
- ▷ Georeferencing imagery
- ▷ Creating paths and routes
- ▷ Animating and narrating tours
- ▷ Converting Shapefile to KML
- ▷ Mapping with Google Fusion Tables

CHAPTERS 1 THROUGH 4 INTRODUCED READERS to the features and applications of Google Earth. This chapter provides users with hands-on tutorials that will guide users through specific functions and tasks in Google Earth and related applications. After becoming comfortable with these tutorials, readers will be able to create, find, convert, and map geospatial data using Google Earth. This knowledge can then be integrated into the classroom, in workshops, and in training sessions.

Tutorial: Interactive Mapping in Google Earth

This tutorial introduces users to creating dynamic placemarks (points of interest), georeferencing, as well as the importing images and KML files. You can also learn how to create a walking tour, and work with screen overlays. After the completion of this tutorial, users will be able to create a customized interactive map and share it with other Google Earth users.

Creating a Placemark

Placemarks are one of the simplest yet most versatile tools in Google Earth. In their simplest form they are used to denote a specific location geographically. They can also contain information about this point. For example, they can be used to show where a hotel is located. The name of the placemark could be "Hotel" and its address, accommodations, continental breakfast hours, photos, video tour, and floor plan information can be displayed when the placemark is selected. Placemarks have a variety of uses in Google Earth but essentially represent Points of Interest that users can easily refer back to. Saving a placemark, like a position on the map, is similar to creating a bookmark in a web browser.

A place can be "marked" once it is viewed on the screen. As learned in chapter 1, a place of interest can be found by typing in the name of the location or attraction into the Search box. Locations can also be discovered by typing in latitude and longitude information.

To start, in the search box at the top left of the screen, enter **Mount Mabu, Mozambique** and click on Search. Google Earth will zoom into the Mount Mabu rainforest. There will be a placemark present that Google Earth marked; however, this placemark cannot be customized. A new one will need to be added by the user and placed over the existing one. To do this, select the placemark button in the toolbar. This will open a new window which hosts various properties of the placemark, all of which can be modified.

To take advantage of the customizability of Google Placemarks, additional information will be added into this pop-up window. First, enter a desired name for the placemark in the Name field. In the Description folder, which is the default folder, users will see a blank page which is HTML friendly. This is the section that will make the placemark dynamic. Users can add a description of the placemark using a variety of text fonts, sizes, and colors. Users can add links to webpages, to videos, and embed images and photographs. Tables and frames can also be added, including other elements that can be seen in web browsers, including APIs and widgets. Using the "Add image …" and "Add link …" buttons will automatically enter the HTML. Otherwise, HTML tags need to be typed into the description box. Alternatively, if users wish to keep the placemark description simple and use regular text and not use HTML code, this is supported as well. To get a sense of how HTML code works in the description box, simply copy the text printed in table 5.1, and paste it into the box.

Once the HTML code is pasted into the Description box, it should look like figure 5.1. Click on OK, and then click on the yellow pushpin to see the results of the code. It should look very similar to figure 5.2.

To make changes in the placemark's properties, or its location on the map, simply right-click on the pushpin to return to the properties menu. While this window is opened, users can move the icon on the map. Take another look at the HTML code to understand what each line of code is displaying. The first line displays the image, denoted by the HTML code <img>, followed by the image size. The second line uses a table <table> to properly align and configure the text for the photo credit. Notice that specific font and color has been noted. The third section provides information about Mount Mabu, and it doesn't require any special code. The last section is another table which provides an image and link to a YouTube video. Users who are new to using HTML will find many resources online that will assist with creating dynamic text.

If using the Add image button to include images, it may appear that only online images can be added. In fact, images from the local computer can also be included, using the path where the image is stored on the computer.

Table 5.1. Dynamic Text for Mount Mabu Placemark

```
<img src="http://www.clubofmozambique.com/solutions1/sectionnews/data/tourism/mount_mabu.jpg" width="460"
height="230">

<table><tr><td width="100%"><p align="right">Photo courtesy of Club of Mozambique</p></td></tr>
<tr><td><font face="Arial" color="#2B116A" size="6"><b>$[name]</b></font>
</td></tr></table><br>

Mount Mabu is a mountain in northern Mozambique, famous for its old-growth rainforest. Mount Mabu is approximately
1,700 metres (5,600 feet) high and the forest covers about 7,000 hectares (27 square miles). Although it was well-known
locally, the Mount Mabu forest and its extremely diverse wildlife were unknown to plant and animal scientists until 2005.
It was "discovered" by scientists from Kew Royal Botanic Gardens by browsing Google Earth's satellite view to look for
potential unknown wildlife hotspots in Africa. It is frequently referred to as the "Google Forest". Wikipedia, 2015.

<table><tr><td width="155">
<p align="center"><a href="https://www.youtube.com/watch?v=ZTMyDtHZXIc"><img src="http://www.fauna-flora.org/
wp-content/uploads/Butterflies-on-Mount-Mabu-Credit-Dr-Julian-Bayliss-FFI.jpg" width="110"></a></p>
</td><td><br>Watch "The Lost Forest of Mount Mabu"</a>, a three-minute video featuring Julian Bayliss.</td></tr></
table>
<br>
```

The yellow pushpin which represents the placemark can be changed to another icon. In the Properties menu, beside where the placemark name was added, there is a pushpin symbol. Clicking on that will open up a large array of icon symbols that can be selected to replace the pushpin. Select the green tree symbol and click OK.

The folder to the right of the Description folder is Style and Color. This refers to the display of the label and the icon. Users can customize the color, scale, and opacity of each. Select a lighter color for the Label, and leave the icon as is.

When satisfied with the placemark, it will need to be saved; otherwise, once Google Earth is shut down, all information will be lost. Whenever a new placemark is created, it will be docked on the left side of the program, found under Places. It is best to keep placemarks organized and to place them into folders. To create a new folder, under Places, right click in the white space away from any of the folders and files. Navigate to new -> folder. It is very important that all placemarks are saved in the same folder when wanting to share the folder with others. The folder can be shared via e-mail or file transfer, but

HOW TO FIND AN IMAGE'S PATH ON A COMPUTER

To locate the exact path of where the image is stored on a local computer, first locate the image and right-click on the file name. In the Properties menu (or Get Info on a Mac), the "Location" information is displayed showing the folder where the image is, but not the actual image itself; therefore, one more step is required. After the folder name, add \ and the name of the image, including the file format. An example of the resultant string of text is: C:\YourName\Pictures\danaporter.jpg.

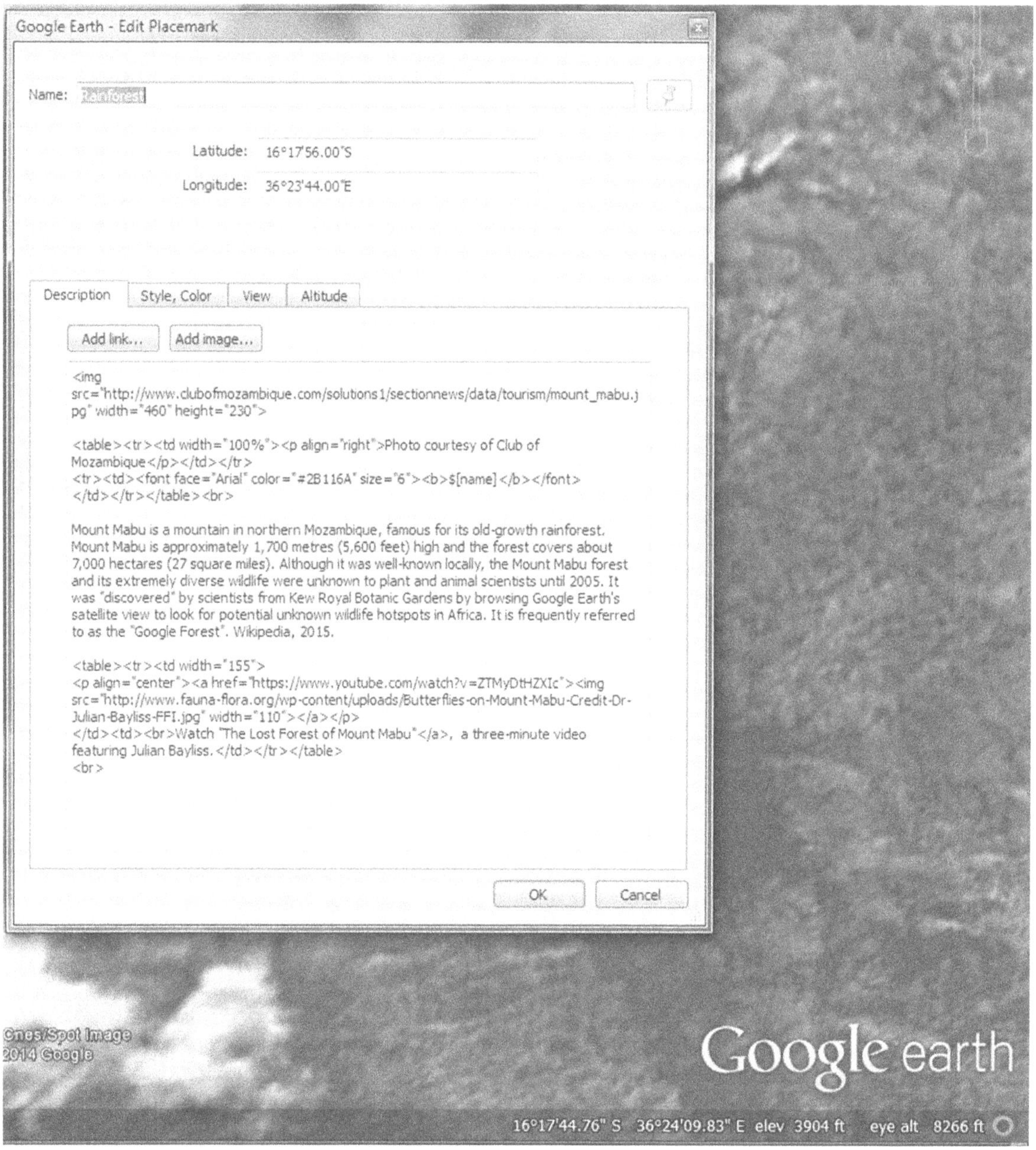

Figure 5.1. HTML code typed into the Description Box.

Source: Google Earth 7.1, <http://earth.google.com/>. Map Data: 2014 Cnes/Spot Image, 2014 Google

first it needs to be saved locally on the computer. Right click on the folder and select **Save Place As**. This will save the placemark as a KML or KMZ file.

Georeferencing Imagery

Besides including images in placemarks, Google Earth also supports adding imagery directly onto the aerial imagery. This differs from adding photos with placemarks, as the placemarks are just one point and so are useful for placing an image; however, they are not useful when trying to compare changes to the terrain over time from historical imagery. In the latter case, the image or map will need to be placed spatially, otherwise known as georeferencing the imagery.

Figure 5.2. Dynamic placemark displayed in Google Earth

Source: Google Earth 7.1, <http://earth.google.com/>. Map Data: 2014 Cnes/Spot Image, 2014 Google; Image Source: Club of Mozambique, <http://www.clubofmozambique.com/solutions1/sectionnews.php?secao=mozambique&id=14652&tipo=one#>

Taking a paper map and converting it into digital form can be very handy. With the digital copy a number of techniques can be used to place the map spatially, offering a variety of uses. Historical maps can be downloaded from a number of online resources, but to make this task simpler, a campus map of the University of Waterloo will be used.

First, the university will need to be zoomed into. In the search box, type University of Waterloo, Waterloo, ON and then move the view extent until Ring Road is clearly in view. If unsure where the road is, clicking on the Roads layer may help.

Next, select the Image Overlay button from the toolbar at the top of the screen, as this will open a new window called the Properties Pane. It looks similar to the placemark window, but it has an additional textbox named "Link." This textbox will be used to tell Google Earth where the image is that needs to be georeferenced. Images can be used directly from the Internet by pasting the URL into this box, or the browse button can be used to navigate to an image on the local computer. For this tutorial the image will be obtained from <http://www.math.uwaterloo.ca/~bguenin/images/campusmap.jpg>. Insert this URL into the link textbox or download the image and browse to its location on the local computer. Selecting OK at this point will display the image at the current extent of the window. Since the image will be placed in the current screen position, it can be helpful to first navigate to approximately the image's correct position to avoid further hassle later on.

The image overlay file will also now pop up on the left side of the screen under My Place, the same way the rest of the features have. The Image Overlay window can be found again by right clicking the image overlay and selecting Properties. When this is opened it unlocks a variety of options for moving the image. There will be green lines along the edges, corners, and middle of the image which can be used to drag the image around to its correct position. The image can be rotated by going to the Location tab and modifying the numbers in the Rotation input box. The Convert LatLonQuad button on this tab will allow the image to be placed using coordinates for each corner of the map if that is easier to perform. See figure 5.3 for the locations of each option. After the image is positioned properly, select OK. The georeferenced image can be saved in the same way as the placemarks are saved.

Paths/Elevation

Similar to the ruler tool, the Path tool allows the user to draw a route (path) and then access information about that path. In this example, elevation changes will be explored along a path in an area with very drastic altitude changes.

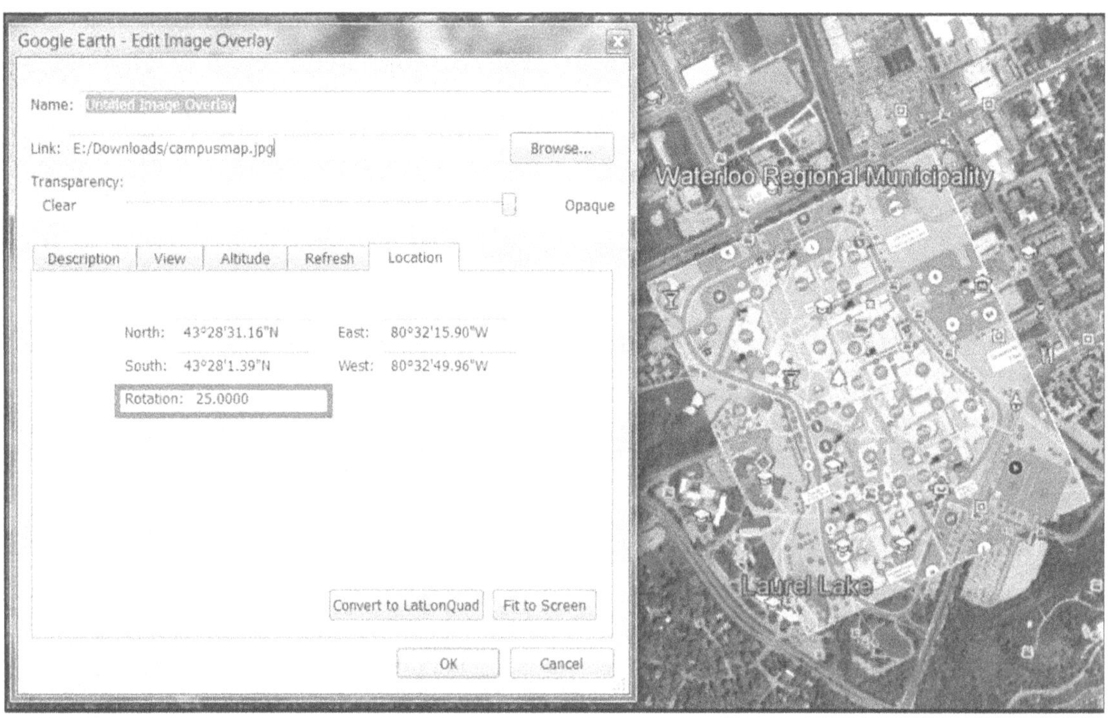

Figure 5.3. Image Overlay Options in Google Earth.

Source: Google Earth7.1, <http://earth.google.com/>. Map Data: 2015 DigitalGlobe, 2015 Google; University of Waterloo Campus Map, https://uwaterloo.ca/map/

Navigate to Mount Everest and select the Path tool, available from the toolbar. A window will appear, activating the tool. While drawing, clicking on the left button of the mouse will add a point, while clicking in the right will delete a point. Holding down the left mouse button will enable freehand drawing. Draw a path up the mountain from the nearby settlement. Once finished, click the OK button on the properties pane. On the left-hand side of the screen, the path will be shown in My Place. Right click on the path and select Show Elevation Profile. A new interface will now be displayed at the bottom of the screen. This pane allows access to different types of information including elevation at any particular point and distance from origin. As the cursor is moved along the graph the point corresponding to the location on the X access will be displayed on Google Earth spatially.

The Path tool is commonly used to create tours, walking/running routes, and trails. Since the distance is recorded of each line segment, this tool is an excellent feature to use to plan out outdoor activities. Paths can also be created by using the Directions feature. A walking path for example can easily be established by connecting two placemarks. Simply right-click on the placemark that is the starting point, and select "Directions from here." Then right-click on the destination placemark and select "Directions to here." Google Earth automatically plots the tour, available for walking, driving, or biking. This path needs to be saved like the other map features. Underneath the directions in the side panel, click "Save route to My Places" and then close the Directions panel. Now drag the newly saved "Route" into the folder that was created earlier.

Narrated and Animated Tours

As discussed in chapter 4, Google Earth offers recording features which capture movement of the screen within the program. For this tutorial, a tour will be created that will record exact navigation in the 3D environment with an audio commentary using a drawn route. Layers that are available in Google Earth can be included in the tour, such as Photos, 360-degree Google Street View, and 3D buildings. This tutorial will be creating a tour of the Forbidden City in Beijing, China. Before creating the tour, placemarks will need to be created of the following locations:

- Tiananmen Square (39°54'26.10"N, 116°23'28.28"E)
- Wenhua Palace (39°54'49.26"N, 116°23'34.91"E)
- Main Halls of the Forbidden City in Beijing (39°54'56.21"N, 116°23'27.03"E)
- Wuying Palace Buildings (39°54'48.32"N, 116°23'17.74"E)

The tour will begin at the Tiananmen Square placemark. Using the Path Tool, draw the following tour paths: Tiananmen Square to Wenhua Palace, Wenhua Palace to Main Halls, Main Halls to Wuying Palace, and Wuying Palace to Main Halls. Under the Layer panel, turn on the Photos and 3D Buildings layers. Then Click the Add Tour button in the toolbar, or go to the Add menu and select Tour. The tour recording tools appear in the lower-left corner of the window.

Click the Record button to start recording the actions and movements in Google Earth. The button changes to red and the time counter begins counting up, showing the duration of the current Tour. There are 3D buildings for the placemarks that have been set up in the 3D viewer. Using the Navigation Controls in the top left corner and the mouse to move, tilt, rotate the globe, zoom in around the 3D building. There are some photo icons surrounding Tiananmen Square. Click one and a pop-up window showing the photo and its description appears.

Once the 360 degree view of the Tiananmen Square is finished, the stop recording button can be pressed. Tour Player appears in the bottom-left corner of the 3D viewer, and the tour can be played once the recording is complete. To save the tour, click the Save to File button in the Tour Player, and give it a name. The new KML file will be saved in "My Places." It can be placed into the folder with all of the other placemarks.

Audio commentary can be recorded once the tour has been created. Adding an audio clip in this way is easy, but the quality of the commentary can be improved if a script is prepared in advance. A microphone must be connected to the computer.

First, the tour recording needs to be played again to figure out how long each segment of the tour takes. When ready, the tour is played and then re-recorded, this time with the microphone button instead of the record button. The recording will start and Google Earth will capture audio from the computer's microphone. The recording can be turned on/off by clicking the microphone button again. Once finished, click the "Save" button in the Record Tour Controls to save the tour. Audio recorded through your microphone is captured as an .mp3 file when saved under "My Places" as a KMZ file.

Converting Shapefile to KML

The possibilities for map making in Google Earth are endless once users learn how to convert popular GIS file formats into Google Earth friendly KML file formats. Because shapefiles, tables, and text files have been around longer than KML files, most libraries collect and manage them over KML files. Until KML file formats become the most popular geospatial format, shapefiles will need to be converted to be used in Google Earth. As was seen in chapter 4, there are numerous applications that offer format conversions. Some are available in desktop GIS programs; others are available freely online. They all differ from one another, as some offer customizability in the way the data are displayed, whereas others will simply convert the geometry and leave some of the pertinent information behind. A couple of different applications will be introduced to readers, guiding them through converting shapefile to KML.

Geocommons

Geocommons <http://geocommons.com> is currently one of the best online geospatial data conversion and mapping service providers. Although the Shp to KML conversion service is free, Geocommons requires registration for an account. First time users will need to click on "Sign up for Free" button, and then follow the instructions to obtain an account.

Once registered and logged in, the top of the page will have two useful buttons, including "Upload Data" and "Your Library." Before data can be uploaded, the user needs to become familiar with the components of a shapefile. Note that a shapefile comprises multiple files with the same name but different extensions. Users must upload all mandatory files in order to ensure correct results. They are:

- {filename}.shp file (mandatory): The main file stores all geographic features;
- {filename}.shx file (mandatory): The spatial index file stores the index for searching;

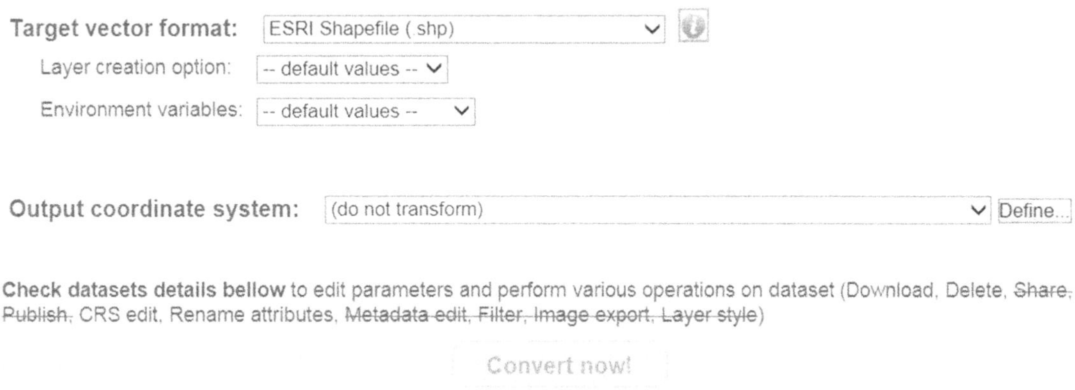

Figure 5.6. MyGeoData conversion options.

the project, under the Dataset Details and Operations section. Once satisfied with the output file format and options, click the "Convert Now" button to begin and complete the conversion process. Once the conversion is complete, save the data and import it into Google Earth to view and edit the file.

Desktop GIS KML Converters

Although less convenient, many desktop GIS programs have a file conversion tool, enabling the importation of shapefiles and the exportation of those same files in KML file format. ArcGIS offers such a tool, but because it is a licensed proprietary product, open source options may be more desirable for libraries. Open source products like Quantum GIS and MapWindow GIS are two products that offer conversion utilities. A brief summary about MapWindow GIS' Shape2Earth plug-in is included because it is a simpler program to work with but offers more advanced options when creating the KML file.

MapWindow is a free, open source GIS software product and can be downloaded from <http://www.mapwindow.org/>. When users start up MapWindow, they are greeted with a default start dialog titled "Welcome to MapWindow!" In this window, there are a few basic options. The one of interest for KML conversion is the "Add data to this project." If this dialog box is inadvertently closed, it can be brought up again by clicking on the Add button on the toolbar.

Navigate to the folder that contains the shapefile that needs to be converted and add it to the map. To select the target layer to convert, go to View > Panels > Legend in the menu bar. Simply click on the layer name in the Layers tab on the Legend panel. When the Shape2Earth menu comes up, there is an option to Export to KML. There are a variety of options to use to customize the Google Earth KML file.

KML Options

In the KML options tab, users can change the default layer name that will appear under the Layers pane in Google Earth. The Properties Tab allows for some additional features and customizability to present spatial data in Google Earth. There are a few different options between point data and polygon data. Users can choose from a wide plethora of point icons, anything from the default yellow pushpin to a rockslide warning. Users can also add their own icon. However, beware that the KML will create a link to the file instead of actually storing it, so while it may look great on your computer screen, it will not appear if opened by someone else on their screen.

The Fill Polygon option will fill the polygon. When unchecked, the polygon will only have the lines showing. Checking off the Create Labels will create a centroid point where a label will be generated.

The GIS Data tab allows users to take the attribute data and assign it to the KML data. Setting the name field provides a specific name for each point, rather than [No name] for all the points. Each attribute selected to write to the KML will appear in the Google Earth pop-up balloon under the KML description.

The Balloon tab allows for customization of the Google Earth data balloon. It offers users some basic information about the KML file pre-loaded into the information balloon.

Creating the KML File

Once the options have been set, and the layer has been chosen, simply click the Google Earth icon button directly below the File option in the toolbar. Browse to a folder that the file will be saved in, and click Save. Now it can be viewed in Google Earth.

Mapping with Google Fusion Tables

Google Fusion Tables is an online data management application designed for users to collaborate, visualize data, and publish their maps online. Google Fusion is often used to convert spreadsheets into maps, as it offers a relatively easy and seamless geocoding functionality. Since the free version of Google Earth does not support geocoding nor the importation of tables, Google Fusion's ability to geocode and convert a spreadsheet into a Google Earth readable format is favored by many users. This tutorial will teach users how to transform location information stored in an Excel spreadsheet into a Google Fusion Table that can be geocoded and mapped directly, or exported as a KML file.

Format and Importation of Excel Spreadsheet

Google Fusion is very specific with its address columns, so specific formatting needs to be followed in order for the operation to be successful. The tutorial offers a pre-made table which has been set up in the correct format. The table is a sample list of Tim Horton's coffee shops in the cities of Kitchener and Waterloo, Ontario. Copy and paste the information from table 5.2 into an Excel spreadsheet.

Review the table to see how the information is formatted. Note that the headings do not have any spaces, and that the address includes the street number, street name, city, province, and country—in that order. A Google account is required to use Google Fusion. Open the homepage of Google Drive <http://www.google.com/drive/apps.html> and log in.

Google Fusion is an application that is a part of Google Drive and can be accessed through the "Create" button. Click "Fusion Table" and browse to the Excel file using the "Import new table" feature. Assign a name for the table and fill out the other essential information it is asking for.

The imported table, as seen in figure 5.7, will be available under the "Rows 1" tab. Note that Google Fusion automatically recognizes the addresses and highlights the locations in yellow. Scroll down and check to make sure all addresses are highlighted.

Table 5.2. A Sample List of Tim Horton's Coffee Shops

NAME	ADDRESS	WIFI	DRIVETHRU
Tim Hortons	109 KING St. N Waterloo Ontario Canada	Yes	Yes
Tim Hortons	94 BRIDGEPORT RD Waterloo Ontario Canada	Yes	No
Grand River Hospital	835 KING ST W Kitchener Ontario Canada	Yes	No
Wilfred Laurier University	200 KING ST N Waterloo Ontario Canada	Yes	No
Wilfred Laurier University	75 UNIVERSITY AVE W Waterloo Ontario Canada	Yes	No
Tim Hortons	638 BELMONT AVE W Kitchener Ontario Canada	Yes	Yes
Tim Hortons	65 UNIVERSITY AVE E Waterloo Ontario Canada	Yes	Yes
Tim Hortons	730 KING ST W Kitchener Ontario Canada	Yes	Yes
University of Waterloo	200 UNIVERSITY AVE W Waterloo Ontario Canada	Yes	No
University of Waterloo	200 UNIVERSITY AVE W Waterloo Ontario Canada	Yes	No
University of Waterloo	200 UNIVERSITY AVE W Waterloo Ontario Canada	Yes	No
Gas Station	151 COLUMBIA ST Waterloo Ontario Canada	No	Yes
Tim Hortons	384 KING ST N Waterloo Ontario Canada	Yes	Yes
Tim Hortons	415 KING ST N Waterloo Ontario Canada	Yes	No
Tim Hortons	227 KING ST S Waterloo Ontario Canada	Yes	No
Tim Hortons	350 WESTMOUNT RD Kitchener Ontario Canada	Yes	No
Tim Hortons	465 PHILIP ST Waterloo Ontario Canada	Yes	Yes
Tim Hortons	7 851 FISHER HALLMAN Kitchener Ontario Canada	Yes	No
Tim Hortons	354 HIGHLAND RD W Kitchener Ontario Canada	Yes	No
Tim Hortons	1500 HIGHLAND RD Kitchener Ontario Canada	No	No

Information in the attribute table can be edited under the Cards 1 tab. They can also be filtered to include or exclude information based on mapping choices and needs. Within this tab there are also options to delete, or create duplicates of specific locations. To change the table view, from the "File" menu. select "Create View." Here you can select the columns that you want to display and assign a name to the table view. For example, you can make a new table omitting WiFi information.

Geocoding and Mapping the Spreadsheet

Once all the desired attribute information is displayed in the spreadsheet, the final step is to run the geocoding processor and visualize the results in a map format. To geocode the Tim Horton's locations, click on the "Map of Addresses" tab, found directly to the right of the "Row 1" and "Cards1" tab. This will start the address geocoding process automatically and may take a few minutes to complete. Each Tim Horton's location will be represented with a point icon on the map. Zooming in and clicking on one of the icons will provide the information stored in the attribute table.

Figure 5.7. Imported table in Google Fusion

With Google Fusion Tables, there is the option of querying the data and then displaying the results on the map. To create a query, click on the "Filter" tab in the top left-hand corner of the map. Set a condition by clicking on the attribute that needs to be queried, for example, "WiFi." This opens a pop-up on the left-hand side of the page. Check off "Yes" for "WiFi" which will only show locations which have this attribute. Immediately, there will be a change in the map—only the Tim Horton's locations with WiFi will be shown.

The icons representing the coffee shops can easily be changed. Under the "Configure map" section to the left of the map, click "Change feature style…" This will open up a pop-up that contains options to change the style of the icon. Under "Points," change the icon from its original color ("small_red") to "small_blue."

The final map can be shared or exported out in KML format. To share the map, click on "Share" on the top right side of the page. In the "Share setting" windows, set "who has access" from "Private" to "Public on the web." To share via e-mail, on the top right side of the page, click "Share" and then copy and paste the link into an e-mail. To export the locations out as a KML file, click "File—Download" in the top left-hand corner of the page. The KML file can then be viewed in Google Earth.

Key Points

Google Earth has many uses and is an attractive option for users who have varying levels of mapping experience. It acts as a platform for discovery, research, mapping, presenting, and for analysis, offering users a number of features that are built-in as well as compatible with external programs. Its customizability offers users the ability to create personalized and tailored placemarks, to upload photographs and other images, and to create personalized narrated animations of their maps. Users who wish to visualize geospatial data can do so by importing external KML files, including those that have been previously converted from shapefile or an Excel file. Google Earth offers students, librarians, instructors, and other researchers the ability to explore GIS and use a mapping program easily and with little training.

Librarians and other educators are encouraged to go through the guides provided in this chapter and to embed them into classrooms and services offered in their libraries. The capabilities of Google Earth are multi-disciplinary and endless as they pertain to learning, sharing information, and mapping features related to the Earth.

- Users are customizing their maps with HTML code and personal images
- Google Earth offers advanced spatial analysis features like georeferencing and geocoding
- KML files can easily be converted from other spatial files like shapefiles
- Many online geospatial resources and applications provide KML formats allowing for easy importing into Google Earth.

Index

About the Authors

Eva (Hobot) Dodsworth is the Geospatial Data Services Librarian at the University of Waterloo library where she specializes in teaching GIS and map-related content to the university community. Eva's interests include historical cartographic research, teaching geoweb applications, and historical GIS. Eva is also a part-time online instructor for a number of library schools, and continuing education organizations where she teaches the use of GIS technology in libraries. Eva has also written *Getting Started with GIS : A LITA Guide*, and has co-written *Discovering and Using Historical Geographic Resources on the Web.*

Andrew Nicholson is the GIS/Data Librarian at the University of Toronto Mississauga. In this role, he collaborates closely with teaching faculty to develop course assignments based around map-related research questions, geospatial literacy, and experiential learning. Almost from the time of Google Earth's launch in 2005, Andrew has been promoting its use in libraries and in the classroom as a research and teaching tool, overseeing its application in courses across the disciplines including Anthropology, Biology, Classical Studies, Forensic Science, Geography, Language Studies, among many others.

www.ingramcontent.com/pod-product-compliance
Ingram Content Group UK Ltd.
Pitfield, Milton Keynes, MK11 3LW, UK
UKHW051932150926
475560UK00005B/365

9 781442 255